Preface

Two decades ago, the term "algebraic K-theory" had not even been born – the subject was still in embryo. A decade later, there was no doubt of existence; it was uniqueness which seemed at risk as seemingly distinct competing definitions tumbled forth. Since then there has been much reconciliation and development. It is now unquestionably a subject in its own right, clearly flourishing.

Yet in certain respects it has grown too fast. The wider mathematical community has found it hard to keep up with the pace, as anyone trying to find the AMS (1980) classification for the subject will appreciate. (There has, however, been significant acknowledgement of its achievements, most notably in the award of a 1978 Fields Medal to one of its exponents, D. Quillen.) In particular, it has become very difficult for "outsiders" (whether established Mathematicians in other areas, or graduate students wishing to enter the fray) to gain access to the material which must be mastered before current research can be studied. Nobody has written an introduction to the subject since its quite early days. A new one is sorely needed. In addition, Quillen's fundamental tool for the most-travelled path to higher algebraic K-theory, the plus-construction, now receives attention from geometric topologists and homotopy theorists. Again, there is a place for a thorough, up-to-date treatment of the basics.

This book is a modest attempt to meet those needs. In 1976 I gave an eight-lecture graduate course at Oxford, much of its content deriving from (a small portion of) a course Quillen delivered at M.I.T. while I was visiting there in 1974–5. The next few years saw a number of developments, so that by 1981 a repeat course at Oxford comprised sixteen lectures. These now constitute the core of the book, although the inevitable inflation has resulted in only one chapter (the eighth) resembling lecture-sized length.

I have in mind as likely reader an algebraist, topologist or algebraic number-theorist requiring first of all a little motivation before confronting the technicalities. The first three chapters therefore emphasize the history of the topic and its relationship to its "older sibling", topological K-theory. In Chapter 4 the real work begins, in the shape of an introduction to the plus-construction. There is, alas, no escaping the reader's need for a little topological familiarity here, but I have tried to minimize this. The pleasant surprise is in how much algebra, in the form of fundamental group-actions on homotopy and homology groups, lies at the heart of this approach. As a result, the discussion (in Chapters 9 and 11) of seemingly topological properties of the topological space $BGLA^+$ is in large part algebraic. For me, an especially satisfying aspect of the treatment is the way in which the same lemma (3.11) in homology of groups is revealed as lying behind both the de-looping of $BGLA^+$ in Chapter 11 and the passage from equivariant A-representations to KA-theory exploited in Chapters 12 and 13.

Throughout I have tried to complement, rather than reproduce, established references. K-theory veterans, as well as novices, should therefore find something here to interest them. By making the plus-construction the central theme of the work, I have automatically ruled out chances of definitiveness: some results which I state for completeness' sake have been established by quite different methods. The reader should be aware that this is only *an* approach to the subject. I hope it encourages him/her also to contemplate others.

The appearance of this book owes much to a number of people. Prominent have been I.M. James at Oxford and G.E.H. Reuter at Imperial College, London, my two Heads of Department over the relevant years. E. Dror (Jerusalem/Oxford) had helpful comments on a draft of much of the work. My erstwhile algebra colleagues at I.C.L. made continual contributions. In particular, M.E. Keating did a marvellous job scrutinizing drafts; I feel sure his efforts have made the presentation more accessible to algebraists. Last (chronologically), Mrs. M. Robertson can take credit for the outstanding *typographical* clarity of what follows.

Oxford

September, 1981. Jon Berrick.

Contents

1 The Whitehead lemma and K_1A

The beginning is predictable enough. . . We consider a ring A. This assumes associativity. In general we shall also suppose that A has a multiplicative identity 1; at the end of this chapter is a discussion of how the theory can be made to accommodate the case when it doesn't. In order to look at the *general linear group* GLA we first study, for $n \geqslant 1$, GL_nA, which is defined as the group of units $(M_nA)^*$ of the ring M_nA of $n \times n$-matrices with entries in A (under the usual matrix addition and multiplication). The inclusion $M_nA \hookrightarrow M_{n+k}A$ is given by matrix direct sum with the $k \times k$-identity matrix I_k; thus $\alpha \mapsto \alpha \oplus I_k = \begin{pmatrix} \alpha & 0 \\ 0 & I_k \end{pmatrix}$.

Notable elements of GL_nA $(n \geqslant 2)$ are the *elementary matrices*, those with (at most) one non-zero off-diagonal entry and 1s right down the diagonal. Thus the elementary matrix e^a_{ij} has $i \neq j$, $a \in A$ and $e^a_{ij} - I_n$ is the zero matrix apart from a in the (i,j)-slot. E_nA is the subgroup of GL_nA they generate:

$$E_nA = \mathrm{gp}\,\{e^a_{ij} \mid a \in A, \quad i \neq j \in \{1, \ldots, n\}\}.$$

The key rules for manipulating these objects are encapsulated in the relations of the *n-th Steinberg group* St_nA which is designed to be an extension of E_nA. Thus it has generators x^a_{ij}, one corresponding to each e^a_{ij}, subject to the relations

$$x^a_{ij}\, x^b_{ij} = x^{a+b}_{ij},$$

and, if $i \neq \ell$ (so that $n \geqslant 3$)

$$[x^a_{ij}, x^b_{k\ell}] = \begin{cases} 1 & j \neq k, \\ x^{ab}_{i\ell} & j = k. \end{cases}$$

Incidentally, although it doesn't affect this formula whichever way round we define our commutators, for the record we shall be using the notation

$$[x, y] = xyx^{-1}y^{-1}.$$

Since the relations between x_{ij}'s are also valid among e_{ij}'s, the function $x^a_{ij} \to e^a_{ij}$ defines a homomorphism φ: $St_nA \to GL_nA$ whose image is E_nA. I haven't bothered to burden φ with the subscript n because of the compatibility of φ with $GL_nA \hookrightarrow GL_{n+p}A$ and the natural homomorphism ι: $St_nA \to St_{n+p}A$ obtained by extending the domain of subscripts. In other words, if we pass to the limit of each directed set of groups and write $GLA = \cup GL_nA$, $EA = \cup E_nA$, $StA = \varinjlim St_nA$, then φ is a homomorphism from StA to GLA with image EA. Now φ carries a huge weight on its shoulders, for it determines, as we shall soon see, both K_1A and K_2A. Close scrutiny is therefore called for. As ever, Z(G) denotes the centre of a group G.

(1.1) LEMMA: *The homomorphism* φ: $St_nA \to GL_nA$ *has as cokernel* (*the set*) GL_nA/E_nA *and as kernel* $Y_n = St_nA \cap \iota^{-1}Z(St_mA)$ $(2 \leqslant n < m)$.

Proof. A matrix (a_{ij}) commutes with e^1_{k1} if and only if $a_{k1} = 0$ and $a_{kk} = a_{11}$. Because the embedding $E_nA \hookrightarrow E_mA$ adjoins 1's down the extension of the diagonal, it follows that

$$E_nA \cap Z(E_mA) = 1,$$

whence

$$St_nA \cap \iota^{-1}Z(St_mA) \leqslant Y_n .$$

To complete the proof it suffices (bearing in mind that $Y_n = St_nA \cap \iota^{-1}Y_m$) to apply the next lemma to show that $Y_m \leqslant Z(St_mA)$.

(1.2) LEMMA: *Suppose a group* G *is the group union of subgroups* G_1, G_2, *each normal in the subgroup generated by the complement of the other. If a normal subgroup* Y *of* G *has* $Y \cap G_i = 1$ $(i = 1, 2)$, *then* Y *is central in* G.

Proof. For i = 1, 2 Y and G_i are normal in gp $(G \setminus G_{3-i})$. So $[Y, G_i] \leqslant Y \cap G_i = 1$; that is, elements of Y commute with all elements of G_1, G_2. Hence they commute with all products of such elements, which is to say, with all elements of G.

We shall choose for our G_1, G_2 the "first row" and "first column" subgroups of St_nA, defined respectively as the subgroups

$$FR_n = gp\{x_{1j}^a \mid a \in A,\ 1 < j \leqslant n\},\ FC_n = gp\{x_{i1}^a \mid a \in A,\ 1 < i \leqslant n\}.$$

Now any two elements x_{1j}^a, x_{1k}^b commute, so that FR_n is abelian, enabling a typical element to be shuffled into the standard form $x_{12}^{a_2} x_{13}^{a_3} \ldots x_{1n}^{a_n}$, which φ maps to

$$e_{12}^{a_2} \ldots e_{1n}^{a_n} = \left(\begin{array}{c|c} 1 & a_2 \ldots a_n \\ \hline 0 & I_{n-1} \end{array}\right).$$

Such a matrix can only be the identity matrix if $a_2 = \ldots = a_n = 0$. Thus $FR_n \cap Y_n = 1$. Meanwhile

$$FR_n \trianglelefteq gp(St_nA \setminus FC_n) = gp\{x_{ij}^a \mid a \in A,\ 1 \leqslant i \leqslant n,\ 1 < j \leqslant n\}$$

because

$$\begin{aligned} & x_{ij}^{-a}\, x_{1k}^{a_k}\, x_{ij}^{a} \\ &= x_{1k}^{a_k} [x_{1k}^{-a_k}, x_{ij}^{-a}] \\ &= \begin{cases} x_{1k}^{a_k} & k \neq i, \\ x_{1k}^{a_k} x_{1j}^{a_k a} & k = i. \end{cases} \end{aligned}$$

The corresponding conditions on FC_n may be proved similarly, or else deduced from the above via the involutions $x_{ij}^a \leftrightarrow x_{ji}^a$ and $(a_{ij}) \leftrightarrow (a_{ij})$ of St_nA and GL_nA respectively, since φ commutes with these involutions. Finally, we have that $St_nA = gp(FR_n, FC_n)$; in fact, because $x_{ij}^a = [x_{i1}^a, x_{1j}^1]$, $gp(St_nA \setminus FC_n) = gp(FR_n, [FC_n, FR_n]) = FR_n[FC_n, FR_n]$ by the normality already demonstrated. Together with the corresponding result on FC_n, this decomposes St_nA as

$$St_nA = FR_n[FC_n, FR_n]FC_n.$$

In summary, the various conditions of (1.2) are satisfied, so that the proof of (1.1) goes through.

For both St_nA and E_nA ($n \geqslant 3$), each generator is expressible as a commutator of other generators, and thus each element as a product of commutators. Recalling that a group equal to its commutator subgroup is termed *perfect*, we therefore record

(1.3) *For* $n \geqslant 3$, St_nA *and* E_nA *are perfect groups.*

This prompts a digression.

PERFECT GROUPS

Equivalent notational ways of expressing the fact that a group P is perfect are

$$P = [P, P] = P^{(1)},$$

or

$$P/[P, P] = P_{ab} = H^1(P; \mathbb{Z}) = 0.$$

Since perfect groups are to have a pivotal ("central"?) role in the development of the theory, we itemise some of their more obvious properties. First note that homomorphisms send commutators to commutators.

(1.4) a) *The homomorphic image of a perfect group is also perfect.*

Indeed, $\varphi St_nA = E_nA$ provides a ready example. An application of (a) is to the collection of subgroups of a given group, G say. The collection of perfect subgroups is therefore closed under automorphisms of G. Further, it is evidently closed under the operation of group union. Hence

(1.4) b) *Any group* G *has a maximal perfect subgroup, the perfect radical* PG, *which is a characteristic subgroup of* G.

Immediate consequences are

(1.4) c) *If* $\phi : G \to H$ *is a homomorphism, then* $\phi PG \leqslant PH$.

(1.4) d) *If* $\phi : G \to H$ *is a homomorphism and* $PH = 1$, *then* $PG \leqslant \operatorname{Ker} \phi$.

The falsity of the converse to (d) is exposed by the (extreme) example

$$G = \langle x_{12}^1 \rangle * \langle x_{13}^1 \rangle * \langle x_{21}^1 \rangle * \langle x_{31}^1 \rangle \twoheadrightarrow H = St_3\mathbb{Z} .$$

Because G (and so any subgroup of G) is free, $PG = 1$, while $PH = H$. The example further shows that the property of having trivial perfect radical is not preserved by epimorphisms and therefore differs, in general, from the property of solubility, to which it is equivalent in the finite case. This situation may be described by

(1.5) G *is soluble if and only if* $PG = 1$ *and for some* i $G^{(i)}$ *is finite.*

Proof. The appropriate definition of solubility is that $G^{(n)}$ (= $[G^{(n-1)}, G^{(n-1)}]$) be trivial for some n. This immediately forces $PG = (PG)^{(n)} \leqslant G^{(n)}$ to be trivial too. Conversely, the finiteness of some $G^{(i)}$ (or equivalently, the finiteness of some $G^{(j)}/Z(G^{(j)})$, since a lemma of Schur asserts that $G^{(j+1)}$ finite follows) forces the derived series of $G^{(i)}$ to terminate after a finite number of steps, n say. But then $G^{(i+n+1)} = G^{(i+n)}$ says that $G^{(i+n)}$ is perfect, whence $G^{(i+n)} \leqslant PG = 1$.

A key question suggested by (1.4) c) (due to resurface later on in (5.11), (6.8)) asks when an epimorphism preserves perfect radicals. Here is a partial answer. (Recall that the n-th centre $\zeta_n G$ of G is defined by $\zeta_0 G = 1$, $\zeta_1 G = Z(G)$, $\zeta_{i+1} G = \{ g \in G \mid [G, \langle g \rangle] \leqslant \zeta_i G \}$, making $\zeta_0 G \leqslant \zeta_1 G \leqslant \ldots$ the upper central series of G.)

(1.6) *The group extension* $\phi : G \twoheadrightarrow H$ *with kernel* $K \leqslant G$ *has* $PH = \phi PG$ *provided either*

(a) ϕ *is split*,

(b) $K \leqslant PG . \zeta_n G$ *for some* n, *or*

(c) $G^{(j)} \leqslant K . PG$ *for some* j.

Proof. In each case it remains, after (1.4) c), only to prove that $PH \leqslant \phi PG$. For (a), if the monomorphism $\psi : H \rightarrowtail G$ splits ϕ, then (1.4) c) again yields that $\psi PH \leqslant PG$, whereupon $PH = \phi\psi PH \leqslant PG$.

(b) Define $L = \phi^{-1}(PH)$. Since for all i $PH = (PH)^{(i)}$, we have that $L = L^{(i)}K$. To use first the case i = 1, $L \leqslant L^{(1)}.PG.\zeta_n G$ implies

$$L^{(n)} \leqslant (L^{(1)}.PG.\zeta_n G)^{(n)} = (L^{(1)} PG)^{(n)} = L^{(n+1)}.PG ,$$

and so therefore (because $PG = (PG)^{(1)}$)

$$L^{(n)}.PG \leqslant L^{(n+1)}.PG = (L^{(n)}.PG)^{(1)} .$$

In other words $L^{(n)}.PG$ is a perfect subgroup of G, whereupon $L^{(n)}.PG \leqslant PG$. Thus $L^{(n)} \leqslant PG$. Then the formula $L = L^{(n)}.K$ reveals that $L \leqslant PG.K$, leaving

$$PH = \phi L \leqslant \phi(PG.K) = \phi PG.$$

(c) One can argue either directly, viz.

$$PH = (PH)^{(j)} \leqslant H^{(j)} \leqslant \phi G^{(j)} \leqslant \phi(K.PG) = \phi PG,$$

or use (1.5), noting that if $G/K.PG$ is soluble, then so is its image $H/\phi PG$.

In particular, (1.6) b) handles all central extensions, as well as those with perfect kernel. An important example of the latter type is worth commemorating, along with its converse (which applies (1.4) d)).

(1.7) *For a perfect normal subgroup* P *of G,* $P(G/P) = 1$ *if and only if* $P = PG$.

End of digression.

We return now to $E_n A$, and look at its relationship to $GL_n A$. Note that if our ring is already a ring of matrices $M_k A$, then we may equate $M_n(M_k A)$ with $M_{nk} A$, so long as we make no claims concerning compatibility with inclusions $M_n A \hookrightarrow M_{n+1} A$, etc.

(1.8) $\qquad E_2(M_n A) \leqslant E_{2n} A$.

Proof. $E_2(M_n A)$ is generated by e^{α}_{12} and its transpose e^{α}_{21}, where α ranges over all matrices $\alpha = (a_{ij})$ of $M_n A$. If we write

$$\alpha = \left(\alpha' \;\middle|\; \begin{matrix} a_{1n} \\ \cdot \\ \cdot \\ \cdot \\ a_{nn} \end{matrix} \right),$$

then

$$e^{\alpha}_{1,2} = \left(\begin{array}{c|c|c} I_n & \alpha' & \begin{matrix} a_{1n} \\ \vdots \\ a_{nn} \end{matrix} \\ \hline 0 & \multicolumn{2}{c}{I_n} \end{array}\right)$$

$$= e^{a_{1n}}_{1,2n} \cdots e^{a_{nn}}_{n,2n} \left(\begin{array}{c|c|c} I_n & \alpha' & \begin{matrix} 0 \\ \vdots \end{matrix} \\ 0 & I_{n-1} & 0 \\ \hline 0 \ldots & 0 & 1 \end{array}\right),$$

which allows us to iterate, to conclude with

$$e^{\alpha}_{12} = \left(\prod_{i=1}^{n} e^{a_{in}}_{i,2n}\right) \cdots \left(\prod_{i=1}^{n} e^{a_{ij}}_{i,n+j}\right) \cdots \left(\prod_{i=1}^{n} e^{a_{il}}_{i,n+1}\right).$$

The inclusion $GL_nA \hookrightarrow GL_{n+k}A$, $\alpha \leftrightarrow \alpha \oplus I_k$, is implicit in the statements of the next lemmas.

(1.9) *If* $\alpha \in GL_nA$, *then*

(a) α *is conjugate in* $GL_{2n}A$ *to* $I_n \oplus \alpha = \begin{pmatrix} I_n & 0 \\ 0 & \alpha \end{pmatrix}$ *by an element of* E_{2n}, *and*

(b) $\alpha \oplus \alpha^{-1} = \begin{pmatrix} \alpha & 0 \\ 0 & \alpha^{-1} \end{pmatrix} \in E_{2n}A.$

Proof. First observe that, for any unit $a \in A^*$,

$$\begin{pmatrix} 0 & a \\ -a^{-1} & 0 \end{pmatrix} = e^{a}_{12}\, e^{-a^{-1}}_{21}\, e^{a}_{12} \in E_2A.$$

Now $\alpha \in (M_nA)^*$. Accordingly $\begin{pmatrix} 0 & \alpha \\ -\alpha^{-1} & 0 \end{pmatrix} \in E_{2n}A$ by (1.8). It remains to observe that

$$\begin{pmatrix} \alpha & 0 \\ 0 & I_n \end{pmatrix} = \begin{pmatrix} 0 & I_n \\ -I_n & 0 \end{pmatrix} \begin{pmatrix} I_n & 0 \\ 0 & \alpha \end{pmatrix} \begin{pmatrix} 0 & -I_n \\ I_n & 0 \end{pmatrix},$$

while

$$\begin{pmatrix} \alpha & 0 \\ 0 & \alpha^{-1} \end{pmatrix} = \begin{pmatrix} 0 & \alpha \\ -\alpha^{-1} & 0 \end{pmatrix} \begin{pmatrix} 0 & -I_n \\ I_n & 0 \end{pmatrix}.$$

It is more useful to have a generalized version of (1.9) a).

(1.10) *If* $\beta_n \in GL_nA$, *then* β_n *is conjugate in* $GL_{2(m+n)}A$ *to* $I_m \oplus \beta_n$ *by an element of* $E_{2(m+n)}A$.

Proof. An immediate application of (1.9) a) gives $(\beta_n \oplus I_m) \oplus (I_m \oplus I_n)$ conjugate to $I_m \oplus I_n \oplus \beta_n \oplus I_m$. Now if $\alpha = \prod_\lambda e_{i_\lambda, j_\lambda}^{a_\lambda} \in E_{2n}A$, then $\prod_\lambda e_{m+i_\lambda, m+j_\lambda}^{a_\lambda} = I_m \oplus \alpha \in E_{m+2n}A$. Combination with a second application of (1.9) a) produces conjugation to $I_m \oplus \beta_n \oplus I_n$ as required.

Now suppose $\alpha_1, \alpha_2 \in GL_nA$. Then

$$[\alpha_1 \oplus I_n, \alpha_2 \oplus I_n] = [\alpha_1, \alpha_2] \oplus I_n = (\alpha_1 \oplus \alpha_1^{-1})(\alpha_2 \oplus \alpha_2^{-1})(\alpha_1^{-1}\alpha_2^{-1} \oplus \alpha_2\alpha_1),$$

an element of $E_{2n}A$ by (1.9). Thus $[GL_nA, GL_nA] \leqslant E_{2n}A$ and we have

$$[E_nA, E_nA] \leqslant [GL_nA, GL_nA] \leqslant E_{2n}A = [E_{2n}A, E_{2n}A],$$

using (1.3). Passage to the limit gives the instrumental

(1.11) Whitehead LEMMA: $[EA, EA] = [GLA, GLA] = EA = P(GLA)$.

In particular we have EA normal, indeed characteristic, in GLA; moreover GLA/EA = Coker φ is an abelian group and is often known as the *Whitehead group* of A (just as, to be honest, are some of its quotient groups). If the name is a little uncertain, there is no doubt whatever about the symbol: let

$$K_1A = GLA/P(GLA) = GLA/EA = GLA_{ab} = H_1(GLA; \mathbb{Z}).$$

Since the constructions GL, E and St are functorial (in fact E and St even convert ring epimorphisms to group epimorphisms), we have K_1 as a covariant functor from the category of rings (still, for the moment, with a 1) and ring homomorphisms to the category of abelian groups and group homomorphisms. This ought to constitute a simplification, for which one should in principle be grateful. We'll therefore look at some of the easier examples. However, first a comment on the group operation in K_1. The multiplication inherited by any quotient of GLA is of course matrix multiplication; in this instance (1.10) offers an alternative description. For it reveals that, given $\alpha_m \in GL_mA$, $\beta_n \in GL_nA$, the multiplication in K_1A

$$(\alpha_m, \beta_n) \mapsto (\alpha_m \oplus I_n) . (\beta_n \oplus I_m)$$

coincides with

$$(\alpha_m, \beta_n) \leftrightarrow (\alpha_m \oplus I_n) . (I_m \oplus \beta_n) = \alpha_m \oplus \beta_n .$$

We now move in two opposite directions (one at a time). First, we specialise to commutative rings, where, by considering Euclidean rings and fields we can do the computations easily enough. Then we generalise to rings without a multiplicative identity.

COMMUTATIVE RINGS

The easiest examples for the purposes of calculation occur when A is commutative. For the determinant homomorphism $GL_n A \to A^*$ is invariant under the inclusion $GL_n A \hookrightarrow GL_{n+1} A$. It may therefore be regarded as being defined on GLA, where its kernel, the *special linear group* SLA, certainly includes every elementary matrix and thus EA. So it induces on the quotient group $K_1 A$ a homomorphism $det : K_1 A \to A^*$. Now *det* has a right inverse given by $A^* = GL_1 A \hookrightarrow GLA \to K_1 A$. If we define $SK_1 A = SLA/EA$ as the kernel of *det* we then have

(1.12) *For* A *commutative, the determinant homomorphism induces*

$$K_1 A = SK_1 A \oplus A^* .$$

We can carry out the calculation in Euclidean rings without further ado. The key property we require of the Euclidean algorithm, a function $\delta : A \setminus \{0\} \to \mathbb{N}$, is that for any two non-zero elements a, b there is a third, q say, with either $a = qb$ or else $\delta(a-qb) < \delta(b)$. Both $\mathbb{Z}$ and polynomial rings over fields obviously satisfy this condition, as do discrete valuation rings (e.g. $\mathbb{Z}_{(p)}$ and power series rings over fields) and indeed, somewhat vacuously, fields themselves.

Consider a typical element of $SK_1 A = SLA/EA$; suppose $\alpha = (a_{ij}) \in SL_n A$ is a coset representative – we attack its final column. (This of course follows best military tradition. The reader will also observe how we compensate for lack of sophisticated weaponry by sheer, dogged persistence.) At most n–1 of the elements a_{in} can vanish (α is, after all, invertible), which makes the integer $\bar{\delta}(\alpha)$, given by

$$\bar{\delta}(\alpha) = (n-1-\#\{i \mid a_{in} = 0\}) . \min\{\delta(a_{in}) \mid a_{in} \neq 0\} ,$$

well-defined and non-negative, attaining the value 0 only when the final column of α contains a single non-zero entry. Now suppose $\delta(a_{tn}) = \min\{\delta(a_{in}) | a_{in} \neq 0\}$. For each $a_{in} \neq 0$ $(i \neq t)$ the algorithm produces $q_i \in A$ allowing $a_{in} - q_i a_{tn}$ to be either zero or have δ-value less than $\delta(a_{tn})$. If $a_{in} = 0$ or $i = t$, then put $q_i = 0$. It follows that the matrix $\alpha' = (\prod_{i=1}^{n} e_{it}^{-q_i})\alpha$ has its last column composed of elements $a'_{in} = a_{in} - q_i a_{tn}$. Accordingly

$$\bar{\delta}(\alpha') < \bar{\delta}(\alpha) .$$

Iteration finally yields a coset representative with just the one non-zero entry in the final column. Should this be in the ith row where $i \neq n$, then premultiplication by $e_{in}^{-1} e_{ni}^{1}$ converts it to the (n,n)-position. Again, this element emerges as a factor in the expansion of the determinant and so must be a unit, say $a \in A^*$. Use of (1.9) enables us to premultiply by the matrix $I_{n-2} \oplus a \oplus a^{-1} \in E_n A$, to obtain a coset representative $\bar{\alpha}$ with last column comprising $\bar{a}_{in} = \delta_{in}$. Thus

$$\bar{\alpha} = \begin{pmatrix} \beta & 0 \\ \bar{a} & 1 \end{pmatrix} = \begin{pmatrix} I_{n-1} & 0 \\ \bar{a}\beta^{-1} & 1 \end{pmatrix} \begin{pmatrix} \beta & 0 \\ 0 & 1 \end{pmatrix} .$$

Since $\begin{pmatrix} I_{n-1} & 0 \\ \bar{a}\beta^{-1} & 1 \end{pmatrix} \in E_n A$ expands as $\prod_{j<n} e_{nj}^{c_j}$ for suitable elements $c_j \in A$, we are finally left with a coset representative of form $\beta \oplus I_1 \in SL_{n-1}A$. Well, not quite finally, for we now have to repeat the whole procedure to derive a coset representative in $SL_{n-2}A$ and so on, until ultimately in $SL_1 A$ where it can only be the identity matrix. The conclusion is that $SK_1 A = 0$, making $K_1 A \cong A^*$.

PSEUDO-RINGS

To finish this chapter, we fulfil our pledge to discuss the case of rings which may lack an identity element. Let Λ be such a pseudo-ring. Define $m\Lambda$ to be the set $\Lambda_{fs}^{\mathbb{N}\times\mathbb{N}}$ of functions $\alpha : \mathbb{N}\times\mathbb{N} \to \Lambda$ with finite support (in other words, of finite matrices over Λ), on which we impose the obvious (matrix) operations + and $\circ$, given by

$$(\alpha + \beta)(i, j) = \alpha(i, j) + \beta(i, j)$$

$$(\alpha \circ \beta)(i, j) = \sum_{k \in \mathbb{N}} \alpha(i, k)\beta(k, j).$$

(Because $\alpha, \beta \in \Lambda_{fs}^{\mathbb{N} \times \mathbb{N}}$, the summation is finite.) Evidently $m\Lambda$ is another pseudo-ring and m is in fact a functor from the category $\mathbb{R}$ng of pseudo-rings and their homomorphisms to itself. Tempting though it is to iterate this procedure, the following lemma stops the impulse getting out of hand.

(1.13) *A bijection* $\gamma : \mathbb{N} \to \mathbb{N} \times \mathbb{N}$ *induces a natural equivalence* $\gamma^* : m \circ m \overset{\cong}{\to} m$ *of functors* $\mathbb{R}\text{ng} \to \mathbb{R}\text{ng}$.

Proof. Let $\mathbb{N}_1, \mathbb{N}_2, \mathbb{N}'_1, \mathbb{N}'_2, \mathbb{N}''_1, \mathbb{N}''_2$ represent various copies of the natural numbers. Then, in terms of sets, the bijections $\gamma : \mathbb{N}_i \to \mathbb{N}'_i \times \mathbb{N}''_i$ $(i = 1, 2)$ induce natural bijections

$$\begin{aligned} m(m\Lambda) &= (\Lambda_{fs}^{\mathbb{N}'_1 \times \mathbb{N}'_2})_{fs}^{\mathbb{N}''_1 \times \mathbb{N}''_2} \\ &\cong \Lambda_{fs}^{\mathbb{N}'_1 \times \mathbb{N}'_2 \times \mathbb{N}''_1 \times \mathbb{N}''_2} \\ &\cong \Lambda_{fs}^{(\mathbb{N}'_1 \times \mathbb{N}''_1) \times (\mathbb{N}'_2 \times \mathbb{N}''_2)} \\ &\overset{\cong}{\to} \Lambda_{fs}^{\mathbb{N}_1 \times \mathbb{N}_2} = m\Lambda. \end{aligned}$$

This means that if $\gamma(i) = (\gamma'(i), \gamma''(i)) \in \mathbb{N}' \times \mathbb{N}''$, then (for $\beta \in m(m\Lambda)$)

$$\gamma^*(\beta)(i, j) = (\beta(\gamma''(i), \gamma''(j)))(\gamma'(i), \gamma'(j)).$$

It is easy enough to see that γ^* is an additive homomorphism. Verification that γ^* preserves the $\circ$ operation is a little trickier. For $\beta_1, \beta_2 \in m(m\Lambda)$ we can write

$$\begin{aligned} \gamma^*(\beta_1 \circ \beta_2)(i, j) &= ((\beta_1 \circ \beta_2)(\gamma''(i), \gamma''(j)))(\gamma'(i), \gamma'(j)) \\ &= \sum_{k \in \mathbb{N}} (\beta_1(\gamma''(i), k) \circ \beta_2(k, \gamma''(j)))(\gamma'(i), \gamma'(j)) \\ &= \sum_{h,k \in \mathbb{N}} (\beta_1(\gamma''(i), k))(\gamma'(i), h) \,.\, (\beta_2(k, \gamma''(j)))(h, \gamma'(j)) \end{aligned}$$

$$= \sum_{(h,k)\in \mathbb{N}\times\mathbb{N}} (\beta_1(\gamma''(i), \gamma''\gamma^{-1}(h,k)))(\gamma'(i), \gamma'\gamma^{-1}(h,k)).$$
$$. (\beta_2(\gamma''\gamma^{-1}(h,k), \gamma''(j)))(\gamma'\gamma^{-1}(h,k), \gamma'(j))$$

$$= \sum_{n\in N} (\beta_1(\gamma''(i), \gamma''(n)))(\gamma'(i), \gamma'(n)) . (\beta_2(\gamma''(n), \gamma''(j)))(\gamma'(n), \gamma'(j))$$

$$= \sum_{n\in \mathbb{N}} \gamma^*(\beta_1)(i, n) . \gamma^*(\beta_2)(n, j)$$

$$= (\gamma^*(\beta_1) \circ \gamma^*(\beta_2))(i, j).$$

Hence γ^* is a pseudo-ring homomorphism and the proof is complete.

The bijection γ is destined to have its really big moment in another ten chapters' time. For the present, though, it has served its purpose. We now call upon two further functors. The former takes $\mathbb{R}$ng to the category $\mathbb{M}$onoid of monoids and their homomorphisms. It retains the underlying set but synthesises from the two pseudo-ring operations a monoid multiplication ☆, namely

$$\alpha ☆ \beta = \alpha + \beta + \alpha \circ \beta ,$$

for which the previous additive identity 0 serves as the multiplicative identity. In the case of the matrix pseudo-ring/monoid mΛ we next apply the functor which restricts to (mΛ)*, the group of units of mΛ considered as a *monoid.* This is what we wish to regard as the appropriate generalization of GLA; so let us check its credentials. If $1 \in A$ then there is an obvious bijection

$$\text{GLA} \to (\text{mA})^*, \quad \alpha \leftrightarrow \alpha - I .$$

Because $\alpha\beta - I = (\alpha - I) + (\beta - I) + (\alpha - I) \circ (\beta - I)$, this defines an isomorphism, making (mΛ)* a bona fide generalization after all.

Collating all this information gives a valuable corollary to (1.13).

(1.14) PROPOSITION: *The functors* GL *and* GL∘m : $\mathbb{R}$ing → $\mathbb{G}$roup *are naturally equivalent.*

Perhaps the most important way in which pseudo-rings tend to arise is as ideals of rings. The reader has between now and Chapter 3 to decide on the most suitable ring of which mA (A a ring) is an ideal.

2 K_1A and $K^{-1}X$; K_0A and K^0X

This chapter is by way of an historical note, in which we comment briefly on how algebraic K-theory came by its name. That is, we shall see how algebraic K_1, K_0 resemble topological K^{-1}, K^0. (Further details may generally be found in Bass' book [3].) When in the next chapter we come to define K_q $(q < 0)$ and K_2, we shall then be in a good position to discuss those aspects of topological K-theory which the developing algebraic theory might fairly be expected to mimic.

First though, we ought to attempt an answer to the (dis?)ingenuous reader who queries why there should be any relationship between the algebraic and topological theories. To do this, we recall two basic constructions. For k the field $\mathbb{R}$ or $\mathbb{C}$ of real or complex numbers, the set $k(X)$ of continuous k-valued functions on a topological space X acquires a commutative ring structure from that of k, namely straightforward addition and multiplication of values. Provided X is reasonably well-behaved, it is possible to recapture X from $k(X)$. Specifically, the *maximal spectrum* Max(A) of a commutative ring A comprises the set of maximal ideals of A endowed with the Zariski topology: closed sets are the collections of all maximal ideals which contain a given ideal of A. Then the (non-functorial) construction Max(·) is (often) left-inverse to $k(\cdot)$, the point $x \in X$ reappearing in $\mathrm{Max}(k(X))$ as the maximal ideal of functions which vanish at x.

(2.1) *If* X *is a compact Hausdorff space, then* $\mathrm{Max}(k(X))$ *is homeomorphic to* X.

Thus one can impose quite natural topological conditions without strain. The tension comes when we try to impose useful conditions (e.g. ascending chain conditions) on a ring A but at the same time insist that its maximal spectrum be topologically respectable (e.g. Hausdorff). Thus A Noetherian implies Max(A) is a *Noetherian space* (which means that its open sets satisfy an ascending chain condition). On the other hand it is easy to see that all points of Max(A) are closed sets. By considering a descending sequence of closed neighbourhoods of a point, one checks that if X is Noetherian, then $k(X)$ must be the direct sum of as many copies of k as X has connected components. If X is to be Noetherian, then this number must be finite. Now if X is also Hausdorff (hence compact Hausdorff) we may apply Urysohn's Lemma to the effect that k-valued functions on X separate points (or, which amounts to the same thing, (2.1)) to deduce that X is a discrete space composed of only finitely many points. So when X = Max(A) we conclude . . .

(2.2) *A commutative Noetherian ring has a Hausdorff maximal spectrum if and only if it is semi-local* (= *has a finite number of maximal ideals*).

Other ways of expressing this are that the ring is Artin, or that its (Krull) dimension is 0, *dimension* being defined as the supremum of lengths of chains of prime ideals. Thus $dim(\mathbb{Z}) = 1$, while for an indeterminate t, $dim(A[t]) = 1 + dim(A)$ if A is Noetherian. Moreover *dim* is not increased by either localization or passage to quotient rings. This all suggests that algebraists feel most at home when discussing rings of low dimension (= 0 or 1, to be precise). Since few topologists suffer such inhibitions, the resemblance between the following two "stability" theorems is all the more striking. (The concept of dimension via chain length extends to Max(A). Each chain of prime ideals in A determines a chain of closed sets in Max(A), of smaller length whenever distinct prime ideals are contained in precisely the same set of maximal ideals. Now $dim(\mathrm{Max}(A))$ is the supremum of all prime-ideal-determined chain lengths, so we have $dim(\mathrm{Max}(A)) \leqslant dim(A)$. For example, if A is Noetherian then $dim(\mathrm{Max}(A[t])) = dim(A[t])$, and $dim(\mathrm{Max}(\mathbb{Z})) = dim(\mathbb{Z})$, but if A is also a local ring with unique maximal ideal m, then $dim(A) > dim(\mathrm{Max}(A)) = 0$ so long as $m^2 \neq m$.)

(2.3) THEOREM: *If* A *is a commutative ring whose maximal spectrum is a Noetherian space of dimension* d, *then*

$$SL_n A/E_n A \to SK_1 A$$

is surjective whenever $n \geqslant d+1$ *and bijective whenever* $n \geqslant d+2$.

In order to make much of the topological counterpart to this theorem, we need to have a description of K_k^{-1} of a (compact, Hausdorff) space X. Although the standard definition is as the Grothendieck group of k-vector bundles over the suspension space SX, more helpful here is the isomorphism

$$K_k^{-1}X \cong \varinjlim\ [X, GL_n k].$$

Here the set $[X, GL_n k]$ of homotopy classes of maps from X to $GL_n k$ inherits the group structure of $GL_n k$ (*i.e.* matrix multiplication). However, because of the homotopy

$$f_t : GL_n k \times GL_n k \to GL_{2n} k$$

$$(\alpha, \beta) \mapsto (\alpha \oplus I_n) \cdot \gamma_t^{-1} \cdot (I_n \oplus \beta) \cdot \gamma_t ,$$

where

$$\gamma_t = \begin{pmatrix} (\cos t\pi/2)I_n & (-\sin t\pi/2)I_n \\ (\sin t\pi/2)I_n & (\cos t\pi/2)I_n \end{pmatrix},$$

from $f_0 : (\alpha, \beta) \mapsto \alpha \oplus \beta$ to $f_1 : (\alpha, \beta) \mapsto \alpha\beta \oplus I_n$, the group operation could just as easily be regarded as given by direct sums. (This is all-too-reminiscent of Chapter 1.)

(2.4) THEOREM: *If* X *is a compact CW-complex of dimension* d, *then*

$$[X, GL_n k] \to K_k^{-1} X$$

is surjective whenever $n \geqslant d+1$ *and bijective whenever* $n \geqslant d+2$.

(Here, the dimension of X refers to the highest dimension of any of the (finitely many) cells which make up X.) Although I have de-mystified the comparison a great deal by not expressing (2.4) in its usual form in terms of vector bundles, it still looks almost too good to be true. Fortunately it is easy enough to make the similarity plausible, in the following manner.

First of all, the determinant homomorphism $GLk \to k^*$ gives rise to a splitting à la (1.12).

$$[X, GL_n k] = [X, SL_n k] \oplus [X, k^*].$$

The latter summand is well understood. For, homotopically speaking, $\mathbb{R}^* = S^0$ while $\mathbb{C}^* \simeq S^1 = B\mathbb{Z}$. Thus $[X, \mathbb{R}^*]$ may be identified with $H^0(X; \mathbb{Z}/2)$ and $[X, \mathbb{C}^*]$ with $H^1(X; \mathbb{Z})$, so that in the limit we have

$$(2.5) \qquad K_k^{-1}X = SK_k^{-1}X \oplus \begin{cases} H^0(X; \mathbb{Z}/2) & k = \mathbb{R}, \\ H^1(X; \mathbb{Z}) & k = \mathbb{C}. \end{cases}$$

(Topologists will recognise the cohomology group as being the host for the first characteristic class (Stiefel-Whitney when $k = \mathbb{R}$, Chern when $k = \mathbb{C}$) of a virtual k-vector bundle over SX.)

Next, what does $[X, SL_n k]$ look like? Well, it can first be thought of as $\pi_0(\mathrm{Map}(X, SL_n k))$, the set of path-components of the space of continuous maps from X to $SL_n k$. Each such map is in reality an $n \times n$-matrix of maps, and may thus equally be viewed as an element of $SL_n k(X)$ (whose inverse is the matrix of maps corresponding to the map to the inverse matrix in $SL_n k$). This simply exploits the correspondence between Map $(X, \mathrm{Hom}(\mathbb{N}\times\mathbb{N}, k))$ and $\mathrm{Hom}(\mathbb{N}\times\mathbb{N}, \mathrm{Map}(X, k))$. Because $SL_n k(X)$ is a topological group there is further exploitation afoot: we have $\pi_0(SL_n k(X)) = SL_n k(X)/(SL_n k(X))^0$, which represents a simplification inasmuch as one understands the path-component $(SL_n k(X))^0$ of the identity matrix. Here then is the heart of the matter.

(2.6) LEMMA: *For* $n \geqslant 2$, $E_n k(X) = (SL_n k(X))^0$.

Proof. Certainly $I_n \in E_n k(X) \leqslant SL_n k(X)$. However generators of $E_n k(X)$ have the form e_{ij}^f where $f : X \to k$. Any such is connected to $I_n = e_{ij}^0$ by means of the path which at time t $(0 \leqslant t \leqslant 1)$ gives the matrix e_{ij}^{tf} (where $(tf)(x) = t \,.\, f(x)$, $x \in X$). This shows that $E_n k(X) \leqslant (SL_n k(X))^0$. For the converse, since $(SL_n k(X))^0$ is generated by any neighbourhood of the identity, it must be sufficient to establish that every (sufficiently small) neighbourhood of I_n lies in $E_n k(X)$. This formulation is especially amenable to non-standard analysis; so work in some enlargement $^*SL_n k(X)$,where we verify that if $\alpha = (a_{ij})$ is infinitesimally close $(\sim)$ to I_n (*i.e.* $a_{ij} \sim \delta_{ij}$), then $\alpha \in {}^*E_n k(X)$. We argue by induction. (Note that $a = a_{11}$ is invertible because $a_{11} \sim 1$.) The factorization

$$\alpha = \left(\begin{array}{c|c} 1 & 0 \\ \hline a_{21}a^{-1} & \\ \vdots & I_{n-1} \\ a_{n1}a^{-1} & \end{array}\right)\left(\begin{array}{cc|c} a & 0 & 0 \\ 0 & a^{-1} & 0 \\ \hline 0 & 0 & I_{n-2} \end{array}\right)\left(\begin{array}{c|c} 1 & 0 \\ \hline 0 & \beta \end{array}\right)\left(\begin{array}{c|c} 1 & a^{-1}a_{12} \ldots a^{-1}a_{1n} \\ \hline 0 & I_{n-1} \end{array}\right)$$

yields four matrices $\sim I_n$ ($I_1 \oplus \beta$ is, because the others clearly are). The first and last matrices are products of elementary matrices familiar from (1.1), while for the second we appeal to (1.9)b). For the third matrix we can use induction because $\beta \sim I_{n-1}$ and since in proving (1.10) we showed that $I_1 \oplus E_{n-1}A \leqslant E_nA$. So the induction goes through.

The K_1-groups developed in the previous chapter now unfold as a generalization of those already known to topologists (but, ironically, originally invented by algebraic geometers).

(2.7) THEOREM: *If* X *is a compact Hausdorff space, then*

$$SK_1 k(X) \cong SK_k^{-1} X.$$

We can also restate Theorem 2.4 in a form more obviously comparable with Theorem 2.3.

(2.8) COROLLARY: *If* X *is a compact CW-complex of dimension* d, *then*

$$SL_n k(X)/E_n k(X) \to SK_1 k(X)$$

is surjective whenever $n \geqslant d+1$ *and bijective whenever* $n \geqslant d+2$.

Note that by (2.1) $k(X)$ has maximal spectrum X, which is a Noetherian space only in the case d = 0. So although the topological stability theorem can be expressed algebraically, it does not yet have an algebraic proof. Nevertheless we have had plenty of encouragement to extend other topological K-theoretic properties to an algebraic environment.

K_0A

We have seen how, although the group itself had already been an object of study for some time, it was only the recognition of its resemblance to a previously named topological object which led to its being called K_1. At the same time the same process was at work with K_0. Again, the relationship is quite striking, and well worth noting, not so much out of whimsy but because of the inspirational influence of the comparison on the whole subsequent development of the subject.

Recall, in the same notation as previously, that $K_k^0 X$ is defined to be the *Grothendieck group* generated by all isomorphism classes of (finite-dimensional) k-vector bundles V over X subject only to the relations $[V_1 \oplus V_2] = [V_1] + [V_2]$. (Note that by choice of Hermitian form on V one can split any short exact sequence $V_1 \rightarrowtail V \twoheadrightarrow V_2$, *i.e.* $[V] = [V_1 \oplus V_2]$.) Now k-vector bundles over X and their morphisms form a small, abelian category $\mathbb{V}\mathrm{ect}_X$; general arguments then dictate the existence of an embedding into some suitable category of modules. It is not difficult to visualize the image of such an embedding. In the simplest possible case, where X is contractible, all k-vector bundles are (up to isomorphism) just products of X with a finite dimensional k-vector space (= finitely generated, free k-module). In general, finite-dimensionality of vector bundles will correspond to finite generation of modules. Also, the space X, while no longer contractible, is at least composed of contractible patches. Over any patch, U say, the bundle $V|_U$ looks like $U \times k^n$. Its set of sections $\Gamma(V|_U)$ is therefore an n-dimensional k-vector space. A partition-of-unity piecing argument extends any section over U to a section over all of X, leading to a bundle epimorphism from the trivial bundle $X \times \Gamma(V)$ to V. As noted above, this splits, leaving the original bundle V as a direct summand of a trivial bundle. This suggests that our modules also be summands of "trivial" (= free) ones, *i.e.* projective. In fact, by observing that $\Gamma(V)$ is a module over the ring $k(X)$ and that

$$\Gamma(V_1 \oplus V_2) = \Gamma(V_1) \oplus \Gamma(V_2),$$

we see that Γ defines a functor from the category $\mathbb{V}\mathrm{ect}_X$ to the category of finitely generated projective $k(X)$-modules, and have reasonable grounds for hoping that Γ is at least faithful.

(2.9) THEOREM: *The functor* Γ *defines a natural equivalence between the categories of k-vector bundles over a compact CW-complex* X *and finitely generated projective modules over* $k(X)$.

An inverse to Γ is easily described. The ring epimorphism $k(X) \twoheadrightarrow k$ defined by evaluation of a function at an arbitrary point x of X has kernel a maximal ideal m_x. One may check that evaluation at x further induces a vector space isomorphism

$$\Gamma(V)/m_x\Gamma(V) \to V_x,$$

where V_x is the fibre of V at x. Thus the homeomorphism $X \cong \mathrm{Max}(k(X))$ extends, to recapture the vector bundle V from the k(X)-module $\Gamma(V)$.

All this makes the Grothendieck group of finitely generated (f.g.) projective (left) A-modules the inevitable choice for K_0A. Again the group itself, in the guise of the *projective class group* $\widetilde{K}_0A = K_0A/\langle[A^1]\rangle$, effectively predates algebraic K-theory.

(2.10) COROLLARY: $K_0k(X) = K_k^0X$, $\widetilde{K}_0k(X) = \widetilde{K}_k^0X$.

For the record, K_0A is the abelian group generated by isomorphism classes of f.g. projective A-modules, subject to all possible relations of the form

$$[P_1 \oplus P_2] = [P_1] + [P_2].$$

Thus P, P′ define the same element of K_0A if and only if for some finite sequence $Q_1, \ldots, Q_m$ of f.g. projectives there is an isomorphism

$$(\bigoplus_{i=1}^{m} Q_i) \oplus P \cong (\bigoplus_{i=1}^{m} Q_i) \oplus P'.$$

In this event P, P′ are said to be *stably isomorphic*, a property clearly equivalent to the existence of a non-negative integer n for which $A^n \oplus P$ and $A^n \oplus P'$ are isomorphic.

Next, a check of functoriality. If $f : A_1 \to A_2$ is a ring homomorphism and M is a module over A, then

$$f_\# M = A_2 \otimes_{A_1} M$$

is a module over A_2 which is f.g., free or projective whenever M is. The last observation is a consequence of

$$f_\# (M \oplus M') = f_\# M \oplus f_\# M'.$$

Moreover $f_{\#}$ preserves isomorphism classes of modules. Hence $f_{\#}$ preserves stable isomorphism classes, and thereby defines a group homomorphism

$$f_* : K_0A_1 \to K_0A_2 \ .$$

The remaining conditions for (covariant) functoriality are easily seen, namely

$$(id_A)_* = id : K_0A \to K_0A \ , \quad (f \circ g)_* = f_* \circ g_* \ .$$

Note that all f.g. projective modules are in fact free when the ring in question is a field, skew field, principal ideal domain, or a local ring. Thus in these cases $\widetilde{K}_0A = 0$.

3 KqA (q<0) and K_2A

This chapter continues the pre- "plus-construction" historical survey (with History taking second place to Mathematics). Again, I refer the reader to the book of Bass [3] for K_q details $(q < 0)$, while that of Milnor [31] fills in much K_2 material. Further references are indicated in Gersten's review [41 pp.1–40]. On the other hand, topological K-theory is discussed at length in Atiyah [1].

An attractive feature of topological K-theory that immediately caught the eye of algebraists was the Mayer–Vietoris sequence. (The definition of $K^n(X)$ when $n < 0$ will be given shortly.)

(3.1) THEOREM: *If* X_1, X_2 *are compact CW-complexes of which* $X_1 \cap X_2$ *is a subcomplex, then there is an exact sequence* $(n \leqslant 0)$

$$\ldots \to K_k^{n-1}(X_1 \cap X_2) \to K_k^n(X_1 \cup X_2) \to K_k^n(X_1) \oplus K_k^n(X_2) \to$$
$$\to K_k^n(X_1 \cap X_2) \to \ldots \to K_k^0(X_1 \cap X_2).$$

As before, what this tells us about the behaviour of the rings $k(X_i)$ serves as a paradigm. From the co-Cartesian square (denoted by the symbol $\ulcorner.$)

$$\begin{array}{ccc} X_1 \cap X_2 & \hookrightarrow & X_2 \\ \downarrow & \ulcorner. & \downarrow \\ X_1 & \hookrightarrow & X_1 \cup X_2 \end{array}$$

we obtain a Cartesian square (indicated by $.\!\lrcorner$)

$$\begin{array}{ccc} k(X_1 \cup X_2) & \xrightarrow{g_2} & k(X_2) \\ \downarrow{g_1} & .\!\lrcorner & \downarrow{f_2} \\ k(X_1) & \xrightarrow[f_1]{} & k(X_1 \cap X_2) \end{array}$$

By point-set topology (in particular, the Tietze extension theorem), one notes that all four homomorphisms are in fact surjective. So the obvious question to ask is, what can one do with a Cartesian square in $\mathbb{R}$ing

$$(3.2) \qquad \begin{array}{ccc} A & \xrightarrow{g_2} & A_2 \\ {\scriptstyle g_1}\downarrow & \lrcorner & \downarrow{\scriptstyle f_2} \\ A_1 & \xrightarrow[f_1]{} & A' \end{array}$$

given, if absolutely necessary, certain surjectivity assumptions on (most naturally) f_1 and/or f_2? This question played a leading role in motivating the search for further algebraic K-groups. We shall see, in (3.5) and (3.8) below, to what extent it proved possible to define K-groups so as to give it a convincing answer.

There is an especially simple device for defining negative(ly indexed) K-groups in topology. It uses a modified suspension functor $S : \mathbb{T}op \to \mathbb{T}op$ that augments a space by a (disjoint) base-point if it does not already have one, and then applies the usual reduced suspension (viz. smash product with a circle). Then one takes $K_k^{-n}X$ to be $\widetilde{K}_k^0 S^n X$. Sure enough, there is an analogous algebraic process: embed the pseudo-ring mA of all finite matrices in the ring CA of all *locally finite* matrices over A (that is, those with only finitely many non-zero entries in each row or column) whose entries constitute a finite subset of A. In this ring mA forms a two-sided ideal, and the *suspension* SA is then defined to be the quotient ring (just as topological suspension is formed from the cone by identifying an embedded copy of the original space with the apex of the cone). Further aspects of this construction will be discussed in Chapter 11. Again we set, for $n > 0$,

$$(3.3)\ a) \qquad K_{-n}A = K_0 S^n A .$$

This creates a source of possible confusion. For in Chapter 2 we compared K_1A with $K_k^{-1}X$ but now find ourselves looking instead for $K_k^1 X$. At least in the complex case the difficulty is easily cleared up. When $k = \mathbb{C}$ the periodicity theorem reveals that $\widetilde{K}_{\mathbb{C}}^0 S^2 X \cong K_{\mathbb{C}}^0 X$. This allows $K_{\mathbb{C}}^n X$ to be defined for all integers n, with $K_{\mathbb{C}}^n X \cong K_{\mathbb{C}}^{n+2} X$ and in particular, reassuringly, $K_{\mathbb{C}}^{-1} X \cong K_{\mathbb{C}}^1 X$. Note too that $K_{\mathbb{C}}^0 X \cong K_{\mathbb{C}}^1 SX$. It was evidently a vital boost to the credibility of the above algebraic suspension functor when the following fact was established.

(3.3) b) $\quad K_0A \cong K_1SA$.

Further stature was acquired by the following means. A reformulation of the periodicity theorem is that, for $n \geqslant 0$,

$$K_{\mathbb{C}}^{-n}(X \times S^1) \cong K_{\mathbb{C}}^{-n}X \oplus K_{\mathbb{C}}^{-n-1}X.$$

Thus $K_{\mathbb{C}}^{-n-1}X$ is the "non-$K_{\mathbb{C}}^{-n}$ part" of $K_{\mathbb{C}}^{-n}(X \times S^1)$. Now consider the "Laurent polynomial ring" $\mathbb{C}(X)[t, t^{-1}]$: we have

$$\mathbb{C}(X) \subseteq \mathbb{C}(X)[t, t^{-1}] \subseteq \mathbb{C}(X \times S^1)$$

where the projection function

$$t : X \times S^1 \to S^1 \hookrightarrow \mathbb{C}$$

is non-zero and thereby invertible. Moreover, under a suitable norm $\mathbb{C}(X \times S^1)$ can be regarded as the completion of $\mathbb{C}(X)[t, t^{-1}]$, so that the more algebraically presented $\mathbb{C}(X)[t, t^{-1}]$ serves as a fair approximation. On the other hand, there is also an embedding (with Kronecker delta notation)

$$A[t, t^{-1}] \hookrightarrow SA ,$$

$$t^r \leftrightarrow (\delta_{i,j+r}) .$$

(While the locally finite matrix $(\delta_{i,j-1})$ is only a left inverse to $(\delta_{i,j+1})$ in CA, it becomes a two-sided inverse modulo finite matrices.) So

$$\Sigma a_r t^r \leftrightarrow \begin{pmatrix} a_0 & a_{-1} & a_{-2} & a_{-3} & \cdot \\ a_1 & a_0 & a_{-1} & a_{-2} & \cdot \\ a_2 & a_1 & a_0 & a_{-1} & \cdot \\ \cdot & \cdot & \cdot & \cdot & \cdot \end{pmatrix}.$$

Based on the above, our hope would be that $K_{-n}A = K_{-n+1}SA$ is a direct summand of $K_{-n+1}A$, the other summand comprising K_{-n+1} terms. The realisation of this hope is often described as "the fundamental theorem".

(3.4) THEOREM: *For* $n \geqslant 0$, *the inclusions*

$$A \hookrightarrow A[t^{\pm 1}] \hookrightarrow A[t, t^{-1}] \hookrightarrow SA$$

induce an exact sequence

$$0 \to K_{-n+1}A \to K_{-n+1}A[t] \oplus K_{-n+1}A[t^{-1}] \to K_{-n+1}A[t, t^{-1}] \to K_{-n}A \to 0$$

whose monomorphism and epimorphism are (naturally) split.

In consequence the Mayer–Vietoris sequence question (among others) is effectively reduced to a problem about K_1 and K_0. One finds . . .

(3.5) THEOREM: *The Cartesian square* (3.2) *induces an exact sequence* $(q \leqslant 1)$

$$K_1A \to \ldots \to K_qA \to K_qA_1 \oplus K_qA_2 \to K_qA' \to K_{q-1}A \to \ldots,$$

provided that either f_1 *or* f_2 *is surjective.*

By way of example, regular rings form an especially tractable class with respect to K_0, etc. Recall that A is said to be (left) *regular* if it is left Noetherian and every f.g. left module over A has a projective resolution of finite length. These properties allow one to replace K_0A by the Grothendieck group of the category of all f.g. A-modules, which does not suffer the constraints applying to projective modules. The class of regular rings includes the coordinate rings of non-singular (hence "regular") affine algebraic varieties and is closed under passage from A both to the polynomial ring $A[t]$ and to the Laurent polynomial ring $A[t, t^{-1}]$. In fact neither procedure changes K_0, and K_1 is unaltered by the former. In conjunction with (3.4), this says . . .

(3.6) *If* A *is left regular, then for* $q = 0, 1$

$$K_qA \to K_qA[t]$$

is an isomorphism, and for $q < 0$

$$K_qA = 0 .$$

K_2A

We now turn to the altogether thornier question of defining higher K-functors K_q, $q \geqslant 2$. Our experience to date suggests three criteria which ought to be met, namely the appropriate extensions of (3.3), (3.4) and (3.5) above. As a first step, we introduce another algebraic K-functor, Milnor's K_2. Recall from Chapter 1 the (natural) homomorphism $\varphi : StA \to GLA$, whose cokernel was deemed to be K_1A. We now define K_2A to be its kernel. Again, this is evidently functorial. Application of (1.1), in which we let first m, then n, increase to infinity, leads to . . .

(3.7) $$K_2A = \mathrm{Ker}[\varphi : StA \to GLA] = Z(StA).$$

A further description in terms of EA will be given in Chapter 9.

To produce an example of a non-trivial K_2-group we consider the continuous epimorphism $\varphi : St\,\mathbb{R} \to E\,\mathbb{R} = SL\,\mathbb{R}$ (from Chapter 1). This lifts to a map $\widetilde{\varphi}$ from $St\,\mathbb{R}$ to the universal covering space $\widetilde{SL}\,\mathbb{R}$, which sends x_{ij}^{λ} to the final point of the unique path in $\widetilde{SL}\,\mathbb{R}$ starting at the identity and covering the path in $SL\,\mathbb{R}$

$$[0, 1] \to SL\,\mathbb{R}, \qquad t \mapsto e_{ij}^{t\lambda} .$$

Now the covering space lifting theorem ensures that the group action on $SL\,\mathbb{R}$ lifts to one on $\widetilde{SL}\,\mathbb{R}$, while the unique lifting property makes $\widetilde{\varphi}$ a homomorphism as well. Further, because φ is a surjection and a covering projection is a local homeomorphism, the image of $\widetilde{\varphi}$ must contain some neighbourhood of the identity in $\widetilde{SL}\,\mathbb{R}$, and thence contains the subgroup it generates, namely the whole of $\widetilde{SL}\,\mathbb{R}$. In particular, if we look at the kernels of the projections to $SL\,\mathbb{R}$ in the commuting diagram

$$\begin{array}{ccccc} K_2\,\mathbb{R} & & & & \pi_1(SL\,\mathbb{R}) \\ & \searrow & & \swarrow & \\ & St\,\mathbb{R} & \xrightarrow{\widetilde{\varphi}} & \widetilde{SL}\,\mathbb{R} & \\ & {\scriptstyle\varphi}\searrow & & \swarrow & \\ & & SL\,\mathbb{R} & & \end{array}$$

then we see that $\widetilde{\phi}$ restricts to an epimorphism

$$K_2\,\mathbb{R} \twoheadrightarrow \pi_1(\mathrm{SL}\,\mathbb{R}).$$

(More generally, there is an epimorphism

$$K_2\,\mathbb{R}(X) \twoheadrightarrow \pi_1(\mathrm{SL}\,\mathbb{R}(X)) \cong \pi_1(\mathrm{GL}\,\mathbb{R}(X))$$

– the isomorphism being due to the fibration

$$\mathrm{SL}\,\mathbb{R}(X) \longrightarrow \mathrm{GL}\,\mathbb{R}(X) \longrightarrow \mathbb{R}(X)^*$$

whose base $\mathbb{R}(X)^*$ has the homotopy type of a discrete space. However, similarly to before $\pi_1(\mathrm{GL}\,\mathbb{R}(X))$ reduces to

$$[SX, \mathrm{GL}\,\mathbb{R}] \cong K_{\mathbb{R}}^{-1}SX = K_{\mathbb{R}}^{-2}X.)$$

Now by (2.4) $\pi_1(\mathrm{SL}\,\mathbb{R}) \cong \pi_1(\mathrm{SL}_3\,\mathbb{R})$, and Gram-Schmidt orthogonalisation makes the special orthogonal group SO_3 a deformation retract of $\mathrm{SL}_3\,\mathbb{R}$. On the other hand, SO_3 is homeomorphic to 3-dimensional real projective space, whose fundamental group is cyclic of order 2. The non-trivial element of $\pi_1(\mathrm{SL}\,\mathbb{R})$ must correspond to a loop in $\mathrm{SL}\,\mathbb{R}$, based at I, whose lifting to $\widetilde{\mathrm{SL}\,\mathbb{R}}$ begins but does not terminate at the identity. A representative is therefore given by the loop

$$t \mapsto (e_{21}^{-t}\, e_{12}^{2t})^4$$

since when $t = 1$ certainly $\begin{pmatrix} 1 & 2 \\ -1 & -1 \end{pmatrix}^4 = I$. However, its inverse image $(x_{21}^{-1}\, x_{12}^{2})^4$ in $K_2\,\mathbb{R}$ evidently comes all the way from $K_2\mathbb{Z}$ wherein it must also have been non-trivial. In fact it can be further shown that the epimorphism

$$K_2\mathbb{Z} \longrightarrow \pi_1(\mathrm{SL}\,\mathbb{R}) = \{1, -1\},$$

$$(x_{21}^{-1}\, x_{12}^{2})^4 \longmapsto -1$$

is an isomorphism.

Next observe that $(x_{21}^{-1}\, x_{12}^{2})^4 \mapsto (x_{21}^{-1})^4 = x_{21}^{-4} = 1$ via the natural map from $K_2\mathbb{Z}$ to $K_2(\mathbb{Z}/2\mathbb{Z})$. This looks ominous for $K_2(\mathbb{Z}/2\mathbb{Z})$. True enough, it turns out that $\mathbb{Z} \to \mathbb{Z}/n\mathbb{Z}$ always induces an epimorphism

$$K_2\mathbb{Z} \twoheadrightarrow K_2(\mathbb{Z}/n\mathbb{Z})$$

which is non-zero if and only if $n \equiv 0 \pmod 4$.

Additionally, there have been extensive computations for K_2 of fields, skew-fields and related rings. We do not pursue them here, as our attention is more on structural matters. I still have to argue the case for the present definition of K_2; the satisfactory relation with the topological $K_{\mathbb{R}}^{-2}$ has already been noted. We now check whether there are extensions of (3.3), (3.4) and (3.5). First, (3.3) does indeed have a part (c) . . .

(3.3) c) $\quad K_1A \cong K_2SA$.

Secondly, the fundamental theorem presents no problem.

(3.4)′ Theorem 3.4 *also holds when* $n = -1$.

Furthermore, the monomorphism $K_2A \to K_2A[t]$ appearing in (3.4)′ is again an isomorphism when A is left regular (cf. (3.6)). This is all very encouraging: if perhaps too much so, then the Mayer–Vietoris result should give the reader pause.

(3.8) THEOREM: *The Cartesian square* (3.2) *induces an exact sequence*

$$K_2A \longrightarrow K_2A_1 \oplus K_2A_2 \longrightarrow K_2A' \longrightarrow K_1A \longrightarrow K_1A_1 \oplus K_1A_2 \longrightarrow \dots$$

provided that both f_1 *and* f_2 *are surjective.*

While this is quite a fair extension of (3.5) it would obviously be preferable not to have to strengthen the weaker condition on f_1, f_2 appearing there. So perhaps (3.8) holds under the weaker hypothesis of (3.5), namely, that only one of f_1, f_2 need be surjective. Alternatively, if this is the best one can do with this definition of K_2, then perhaps it could be modified slightly so as to provide a straight extension of (3.5). The following counter example (due to Swan [39]) dashes both these hopes.

(3.9) EXAMPLE: Let A be a commutative ring. Two distinguished subrings of its ring of 2×2-matrices, M_2A, comprise the diagonal matrices, which we may write as $A \oplus A = \begin{pmatrix} A & 0 \\ 0 & A \end{pmatrix}$, and the upper triangular matrices $UT = \begin{pmatrix} A & A \\ 0 & A \end{pmatrix}$.

The obvious projection

$$p : UT \to A \oplus A, \quad \begin{pmatrix} a & b \\ 0 & d \end{pmatrix} \mapsto \begin{pmatrix} a & 0 \\ 0 & d \end{pmatrix}$$

is clearly a split epimorphism; we pull it back over the diagonal monomorphism

$$d : A \to A \oplus A, \qquad a \mapsto \begin{pmatrix} a & 0 \\ 0 & a \end{pmatrix}.$$

The fibre-product may be identified with the ring $A[\epsilon]/(\epsilon^2)$ of dual numbers over A, which is to say there is a Cartesian square

$$\begin{array}{ccc} A[\epsilon]/(\epsilon^2) & \longrightarrow & UT \\ \downarrow & \lrcorner & \downarrow p \\ A & \xrightarrow[d]{} & A \oplus A \end{array} .$$

We show that

for no functor K_2 *does this Cartesian square induce a sequence as in* (3.8) *which is exact at* $K_1[\epsilon]/(\epsilon^2)$.

Since any functor must convert a left inverse to a left inverse, this amounts to proving that the exact sequence

$$K_1 A[\epsilon]/(\epsilon^2) \longrightarrow K_1 A \oplus K_1 UT \longrightarrow K_1(A \oplus A)$$

cannot begin with a monomorphism. Now A* is a natural direct summand of $K_1 A$ (1.12), and certainly $(A[\epsilon]/(\epsilon^2))^* \to A^*$ is no monomorphism (for example, the former contains $1 + A\epsilon$). Thus it suffices to prove that

$$K_1 UT \longrightarrow K_1(A \oplus A) = K_1 A \oplus K_1 A$$

is an isomorphism. Since $K_1 A = H_1(GLA)$ (trivial integral coefficients to be assumed), this is just a special case of . . .

(3.10) $\pi_* : H_*(GLUT) \longrightarrow H_*(GL(A \oplus A))$ *is an isomorphism.*

Here $\pi : GLUT \longrightarrow GL(A \oplus A)$ is effectively induced by p – "effectively" because after a permutation of the basis elements of A^{2n} we take $GL_n UT$ to consist of invertible matrices

$\begin{pmatrix}\alpha & \beta\\ 0 & \delta\end{pmatrix}$ in $M_2(M_nA)$, with the case $\beta = 0$ yielding $GL_n(A \oplus A) \cong GL_nA \times GL_nA$. As befits the name, GLUT has enough elements to make

$$\pi : \begin{pmatrix}\alpha & \beta\\ 0 & \delta\end{pmatrix} \mapsto \begin{pmatrix}\alpha & 0\\ 0 & \delta\end{pmatrix}$$

an epimorphism, again split by the obvious inclusion map $GL(A \oplus A) \hookrightarrow GLUT$. So π_* has a right inverse and it remains only to check that, regarded as an endomorphism of GLUT, π induces the identity map on $H_*(GLUT)$. This will be an application of the following highly useful lemma (to resurface in Chapter 11), whose proof is deferred until the end of this Chapter.

(3.11) LEMMA: *Let* $\rho : G \to G$ *be an endomorphism of a group* G *which is the direct limit of subgroups* G_λ, $\lambda \in \Lambda$, *such that* $\rho(G_\lambda) \leqslant G_\lambda$. *Suppose that for each* $\lambda \in \Lambda$ *there exist* $\mu = \mu_\lambda \in \Lambda$ (with $\lambda \leqslant \mu$), *an element* $a \in G_\mu$ *and a homomorphism* $\phi : G_\lambda \to C_{G_\mu}(G_\lambda)$ *such that, for all* $g \in G_\lambda$,

$$g \,.\, \phi(g) = a \,.\, \rho(g) \,.\, \phi(g) \,.\, a^{-1} \;.$$

Then

$$\rho_* = id : H_*(G) \to H_*(G).$$

Application to the present situation encounters one problem: the centralizer of GL_nUT in GLUT is too small for our purposes. Therefore consider instead the normal subgroup H of GLUT comprising $\begin{pmatrix}\alpha & \beta\\ 0 & \delta\end{pmatrix}$ with $\alpha = I$. (So GLUT is the semi-direct product $H \rtimes GLA$.) For each n, π restricts to an endomorphism of $H_n = H \cap GL_nUT$. Now define

$$\phi : H_n \longrightarrow C_{H_n}(H_{2n})$$

$$g = \begin{pmatrix}I & \beta\\ 0 & \delta\end{pmatrix} \mapsto \begin{pmatrix}I & 0 & 0 & \beta\\ & I & 0 & 0\\ & & I & 0\\ 0 & & & \delta\end{pmatrix} = (I \oplus I \oplus I \oplus \delta)e^{\beta}_{14} \;,$$

the last notation employing the isomorphism $GL_2(M_{2n}A) \cong GL_4(M_nA)$. Choose too $a = e^{-I}_{43}$.

Then

$$
\begin{aligned}
g \cdot \phi(g) &= (I\oplus I\oplus\delta\oplus I)\, e^{\beta}_{13} \cdot (I\oplus I\oplus I\oplus\delta) e^{\beta}_{14} \\
&= (I\oplus I\oplus\delta\oplus\delta)\, e^{\beta}_{14}\, [e^{-\beta}_{14}, e^{-I}_{43}] \\
&= e^{-I}_{43}\, (I\oplus I\oplus\delta\oplus I) \cdot (I\oplus I\oplus I\oplus\delta)\, e^{\beta}_{14}\, e^{I}_{43} \\
&= a\, \pi(g) \cdot \phi(g)\, a^{-1} \ .
\end{aligned}
$$

It follows from (3.11) that π induces the identity on $H_*(H)$. Meanwhile, π restricts to the identity on the complement GLA of H. Hence π induces the identity automorphism on the E^2-terms $H_p(GLA; H_q(H))$ of the Lyndon-Hochschild-Serre spectral sequence converging to $H_*(GLUT)$ and thence the identity on $H_*(GLUT)$ itself.

One reason I have felt it worthwhile providing details of this construction is that it serves as a good example to bear in mind when over the next few chapters we study in some depth those maps which induce isomorphisms in homology. For future reference, note that $GLUT \cong MA \rtimes GL(A \oplus A)$ where $MA = \varinjlim M_nA$ is an abelian group under matrix addition.

This concludes the chapter, save for the proof of the technical Lemma 3.11 . Faint-hearted readers may prefer to pass directly to Chapter 4, which will make its own demands on their energies.

Proof of (3.11). An informal comment first, before the technicalities commence. Note that ρ only differs from the identity endomorphism, *id*, because of the intrusion of conjugation by a. Now it is well known that inner automorphisms of groups induce the identity map on their homology, so that in homology this distinction disappears. The purpose of ϕ is to make

$$\eta : G_\lambda \times C_{G_\mu}(G_\lambda) \to G_\mu , \qquad (g_0, g_1) \mapsto g_0 \cdot g_1$$

a homomorphism, thereby allowing an inductive argument through the Künneth formula.

Because the homology functor preserves direct limits, it is enough to show that all the inclusions $G_\lambda \hookrightarrow G$ induce the same homology homomorphisms as the maps $\rho : G_\lambda \to G$. (Note that we do not bother to distinguish notationally between ρ and its restrictions.) The hypotheses set up, for each $\lambda \in \Lambda$, a chain

$$G_\lambda = G_{\lambda_0} \leqslant G_{\lambda_1} \leqslant G_{\lambda_2} \leqslant \dots$$

where $\lambda_{n+1} = \mu_{\lambda_n}$ $(n \geqslant 0)$. For my typist's sake, abbreviate to $G_n = G_{\lambda_n}$. So it suffices to establish, by induction on $n \geqslant 1$, that for $0 < q \leqslant n$

$$(\rho_* - id) \circ i_{n*} : H_q(G_0) \to H_q(G_n)$$

is trivial, where $i_n : G_0 \hookrightarrow G_n$. For n = 1 this is immediate from the observation that

$$\rho(g)^{-1}g \in [G_1, G_1] \;=\; \mathrm{Ker}\,[G_1 \to H_1(G_1)]\ .$$

Assume therefore that $(\rho_*{-}id) \circ i_{n-1*}$, and hence

$$(\rho_*{-}id) \circ i_{n*} \;=\; j_* \circ (\rho_*{-}id) \circ i_{n-1*}$$

(where $j : G_{n-1} \hookrightarrow G_n$), is the zero homomorphism on $H_q(G_0)$ if $0 < q < n$. As a result of the simplification this forces on the map of exact Künneth sequences (with i of course standing for i_{n-1})

$$\begin{array}{ccccc}
\bigoplus_{q=0}^{n} H_q(G_0) \otimes H_{n-q}(G_0) & \rightarrowtail & H_n(G_0 \times G_0) & \twoheadrightarrow & \bigoplus_{q=0}^{n-1} \mathrm{Tor}(H_q(G_0), H_{n-q-1}(G_0)) \\
\downarrow & & \downarrow{\scriptstyle (\rho i \times i)_* - (i \times i)_*} & & \downarrow \\
\bigoplus_{q=0}^{n} H_q(G_{n-1}) \otimes H_{n-q}(G_{n-1}) & \rightarrowtail & H_n(G_{n-1} \times G_{n-1}) & \twoheadrightarrow & \bigoplus_{q=0}^{n-1} \mathrm{Tor}(H_q(G_{n-1}), H_{n-q-1}(G_{n-1}))
\end{array}$$

we have that

$$\begin{aligned}
((\rho \circ i \times i)_* - (i \times i)_*) \circ \Delta_{0*} &= (\rho_*{-}id) \circ i_* \otimes 1 \\
&= in_{L*} \circ (\rho_*{-}id) \circ i_*
\end{aligned}$$

where

$$\Delta_k : G_k \to G_k \times G_k\,, \qquad g_k \leftrightarrow (g_k, g_k)\,;$$

$$in_L : G_{n-1} \to G_{n-1} \times G_{n-1}, \qquad g_{n-1} \leftrightarrow (g_{n-1}, 1).$$

(Thus $\eta \circ in_L = j : G_{n-1} \to G_n$.)

If we let $\hat{a}: G_n \to G_n$ denote conjugation by a, then the main assumption states that

$$\eta \circ (id \times \phi) \circ \Delta_{n-1} \;=\; \hat{a} \circ \eta \circ (\rho \times \phi) \circ \Delta_{n-1},$$

from which we deduce that on $H_n(G_0)$

$$
\begin{aligned}
0 &= \eta_* \circ (\rho \times \phi)_* \circ \Delta_{n-1*} \circ i_* - \eta_* \circ (id \times \phi)_* \circ \Delta_{n-1*} \circ i_* \\
&= \eta_* \circ (id \times \phi)_* \circ ((\rho \times id)_* - id) \circ (i \times i)_* \circ \Delta_{0*} \\
&= \eta_* \circ (id \times \phi)_* \circ in_{L*} \circ (\rho_* - id) \circ i_* \\
&= j_* \circ (\rho_* - id) \circ i_* \\
&= (\rho_* - id) \circ i_{n*} \quad .
\end{aligned}
$$

This completes the inductive step.

4 Acyclic maps

This chapter lays the groundwork for a description of the plus-construction in the next. Apart from a continuation of Example 3.9, it contains no mention of rings. Nevertheless, algebraists may be consoled by the strong algebraic content of some of the concepts introduced here; in particular we are interested in when the fundamental group of a given space acts *nilpotently* on the homology groups of another, and in whether a certain subgroup of the fundamental group is *perfect*. Philosophically, this interplay between algebra and topology is one of the attractive features of the theory, but in practice the topological novice can soon strike problems of digestion. I assume therefore that the reader is armed with a copy of Spanier [38] where necessary, and have provided further referencing to some of the more esoteric points in the text, or wherever I have felt that the index of [38] might be insufficient.

We work in the category of pointed spaces and base-point preserving maps. Whenever we can do so, we choose spaces which are homotopy equivalent to a connected CW-complex that contains only a finite number of cells of any given dimension. This decree is unnecessarily severe, but it saves our issuing special ordinances at various stages.

Of course, the homotopy type of such a space X is uniquely determined by its homotopy groups $\pi_*(X)$ [38 (7.6.24)]. (Base-points are a non-variable part of the machinery and therefore dropped from the notation.) The condition that $\widetilde{H}_*(X) = 0$ (homology and cohomology with trivial integer coefficients unless otherwise stated), *i.e.* that X is *acyclic*, is thus weaker than contractibility of X. Indeed all it tells us about the fundamental group $\pi = \pi_1(X)$ is that it satisfies $H_1(\pi) = H_2(\pi) = 0$, where we recall that the former condition means that π is *perfect.* (Both these facts follow from the Serre spectral sequence of the fibration $\widetilde{X} \to X \to B\pi$ corresponding to the universal covering space $\widetilde{X}$ of X (see p. 51). It gives rise to an exact sequence

$$H_2(X) \to H_2(\pi) \to H_0(B\pi; H_1(\widetilde{X})) \to H_1(X) \to H_1(\pi) \to 0,$$

in which $H_1(\widetilde{X})$ is the abelianization of the trivial group $\pi_1(\widetilde{X})$ [38(7.5.5)], making all terms zero. We shall witness the converse in Chapter 8: given π with $H_1(\pi) = H_2(\pi) = 0$, we can construct an acyclic space with π as fundamental group.) Moreover, the G. Whitehead theorem implies that a space X is contractible if and only if X is both acyclic and has trivial fundamental group [38 (7.5.5)].

The key definition is that a map $f : X \to Y$ is *acyclic* if its homotopy fibre F_f is. (Recall that $F_f = X \times_Y PY$ (denoted E_f in [38 p.432]) is given by the pull-back diagram

$$\begin{array}{ccc} F_f & \longrightarrow & PY \\ \downarrow & \lrcorner & \downarrow \pi_Y \\ X & \underset{f}{\longrightarrow} & Y \end{array} ,$$

where PY is the contractible space of paths $\lambda : I = [0, 1] \to Y$ in Y which start at the base point and whose terminal point is given by $\pi_Y(\lambda)$.) This is clearly a homotopy invariant property, which conveniently allows us to regard f as a cellular map between CW-complexes [38 (7.6.18)], and thence even as an inclusion of a subcomplex in a CW-complex [38 (1.4.12)] when the occasion demands.

(4.1) *If either* f *or* g *is a fibration in the Cartesian square*

$$\begin{array}{ccc} Y' \times_Y X & \longrightarrow & X \\ f' \downarrow & \lrcorner & \downarrow f \\ Y' & \underset{g}{\longrightarrow} & Y \end{array}$$

then f *is acyclic if and only if* f' *is.*

Proof. The less obvious case is when g is a fibration. Then from the commuting diagram

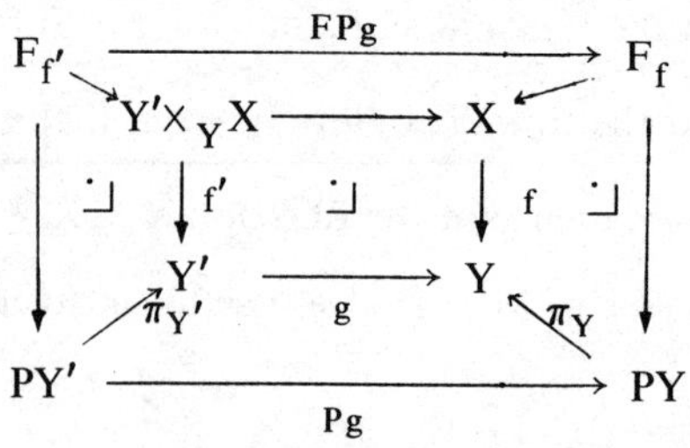

it follows that $g \circ \pi_{Y'}$ and π_Y are fibrations while Pg is a homotopy equivalence. So Pg is in fact a fibre homotopy equivalence over Y [9], as must therefore be its pull-back FPg over X. Whence the result.

Pulling-back by the path-fibration $\pi_B : PB \to B$ gives the following application of (4.1). (Since $F_p = PB \times_B E = (PB \times_B E') \times_{E'} E = F_{p'} \times_{E'} E$, it follows from the above that the homotopy fibres of f and Ff are homotopy equivalent.)

(4.2) *Let*

$$\begin{array}{ccc} E & \xrightarrow{f} & E' \\ & {\scriptstyle p}\searrow \quad \swarrow{\scriptstyle p'} & \\ & B & \end{array}$$

commute. Then f *is acyclic if and only if its induced map on the homotopy fibres of* p, p′ *is.*

(4.3) PROPOSITION: *The following properties of a map* $f : X \to Y$ *are equivalent.*

(i) f *is acyclic*;

(ii) *for any local coefficient system* $\{L\}$ *of abelian groups on* Y,

$$f_* : H_*(X; f^*\{L\}) \to H_*(Y; \{L\})$$

is an isomorphism;

(iii) $$f_* : H_*(X; f^*\{\mathbb{Z}[\pi_1(Y)]\}) \to H_*(Y; \{\mathbb{Z}[\pi_1(Y)]\})$$

is an isomorphism;

(iv) *for* $\widetilde{Y}$ *the universal covering space of* Y, *the map* $f' : X \times_Y \widetilde{Y} \to \widetilde{Y}$ *induced by* f *gives rise to an isomorphism*

$$H_*(f') : H_*(X \times_Y \widetilde{Y}) \to H_*(\widetilde{Y}) .$$

Proof. (i) ⇒ (ii): Since the passage leaves homotopy fibre and homology groups unchanged, pass to the associated fibration [**38** (2.8.9)] of f – also denoted $f : X \to Y$, with fibre F, inclusion $i : F \hookrightarrow X$. Its (slightly generalized) Serre spectral sequence (cf. [**10**]) looks like

$$E^2_{p,q} = H_p(Y; \{H_q(F; i^*f^*L)\}) \Rightarrow H_{p+q}(X; f^*\{L\})$$

where $\{L\}$ is a given, local coefficient system of abelian groups on Y (corresponding to a $\mathbb{Z}[\pi_1(Y)]$-module), so that, $f\circ i$ being nulhomotopic, $i^*f^*\{L\}$ is a trivial local system. Then $\widetilde{H}_*(F; i^*f^*L) = 0$ by the universal coefficient theorem, whence $E^2_{p,q} = \delta_{q0} \cdot H_p(Y;\{L\})$, $E^2_{p,q} = \ldots = E^\infty_{p,q}$ and the edge homomorphism becomes the required isomorphism.

(ii) ⇒ (iii): Trivial.

(iii) ⇒ (iv): Recall there is an isomorphism

$$H_*(Y;\{\mathbb{Z}[\pi_1(Y)]\}) \xrightarrow{\cong} H_*(\widetilde{Y})$$

given as follows (*cf.* [7]). For each $y \in Y$ choose $\widetilde{y} \in \widetilde{Y}$ over it. A singular q-simplex of Y with initial vertex y_0 and coefficient $\omega \in \pi_1(Y, y_0)$ lifts to a unique singular q-simplex of $\widetilde{Y}$, namely the unique lifting with initial vertex $\widetilde{y}_0$ [38 (2.4.5)] composed with the unique covering transformation corresponding to ω [38 (2.6.4)]. Thus $C_*(\widetilde{Y}) \cong C_*(Y) \otimes \mathbb{Z}[\pi_1(Y)]$. This argument extends to give a natural isomorphism

$$H_*(X \times_Y \widetilde{Y}) \cong H_*(X; f^*\{\mathbb{Z}[\pi_1(Y)]\}),$$

making $H_*(f')$ just the composition of group isomorphisms

$$H_*(X \times_Y \widetilde{Y}) \to H_*(X; f^*\{\mathbb{Z}[\pi_1(Y)]\}) \xrightarrow{f_*} H_*(Y;\{\mathbb{Z}[\pi_1(Y)]\}) \to H_*(\widetilde{Y}).$$

(iv) ⇒ (i): In view of (4.1), it suffices to show that f' is acyclic. Because $\widetilde{Y}$ is simply-connected the Serre spectral sequence comparison theorem (4.4) below applies to the map of fibrations

$$\begin{array}{ccc} F_{f'} & \longrightarrow & * \\ \downarrow & & \downarrow \\ X \times_Y \widetilde{Y} & \xrightarrow{f'} & \widetilde{Y} \\ \downarrow{\scriptstyle f'} & & \downarrow{\scriptstyle id} \\ \widetilde{Y} & \xrightarrow[id]{} & \widetilde{Y} \end{array}$$

giving the result immediately.

This list of equivalent formulations is by no means exhaustive. For example, the topologically-inclined reader may like to verify the following.

(4.3) (continued): (v) *the map* f' *suspends to a homotopy equivalence*;

(vi) *for some* $k \geqslant 1$ *the* k*-fold suspension* $S^k f' : S^k(X \times_Y \widetilde{Y}) \to S^k(\widetilde{Y})$ *is a homotopy equivalence*;

(vii)–(x) *as* (i)–(iv) *above, but using cohomology instead of homology*;

(xi) *the homotopy cofibre of* f' *is contractible.*

There is one further, more relevant, reformulation, for which we shall need a definition which takes heed of the importance of fundamental group actions in this theory. This definition is normally given for fibrations but may be taken as referring to maps in general by the usual recourse to the associated mapping-path fibration.

A fibration $F \to E \xrightarrow{p} B$ with *connected* F, E, B is said to be *quasi-nilpotent* if the action of $\pi_1(B)$ on $H_*(F)$ is nilpotent. Although the details of this last term will not be needed, it may be reassuring to have them briefly reviewed.

First, any path $\omega : I \to B$ and singular q-simplex $g : \Delta^q \to p^{-1}(\omega(0))$ determine, by virtue of the homotopy lifting property of p, a map $G : \Delta^q \times I \to E$ over $\omega \circ pr_2 : \Delta^q \times I \to I \to B$ and extending $G_0 = g : \Delta^q \times \{0\} \to E$. If however ω is a loop, then $G_1 : \Delta^q \times \{1\} \to E$ is a q-simplex in $p^{-1}(\omega(1)) = p^{-1}(\omega(0))$ again. Thereby do elements of $\pi_1(B)$ operate on $H_*(F)$. Now given any group π acting on (the left of) another, N say (not necessarily abelian), the semi-direct product $N \rtimes \pi$ operates on N (with N acting on itself by inner automorphisms) and thence its quotients. The fixed-point sets under this action define an *upper central* π*-series* in N:

$$\zeta_0^\pi N = 1, \quad \zeta_{i+1}^\pi N / \zeta_i^\pi N = \mathrm{Fix}_{N \rtimes \pi}(N/\zeta_i^\pi N)$$

$$= \mathrm{Fix}_N(N/\zeta_i^\pi N) \cap \mathrm{Fix}_\pi(N/\zeta_i^\pi N),$$

where of course $\mathrm{Fix}_N(N/\zeta_i^\pi N)$ is just the centre $Z(N/\zeta_i^\pi N)$. If some $\zeta_k^\pi N = N$, then π acts nilpotently on N (of class $\leqslant k$). In particular, if $N \leqslant \pi$, then $\zeta_j^\pi N = N \cap \zeta_j \pi$ is the radical among subgroups of N on which π acts nilpotently (via its inner automorphisms) of class $\leqslant j$. In this case there is also a *lower central* π*-series* $\Gamma_\pi^0 N = N$, $\Gamma_\pi^{i+1} N = [\pi, \Gamma_\pi^i N]$, from which one may define, for an arbitrary group of operators π on N, $\Gamma_\pi^i N = \Gamma_{N \rtimes \pi}^i N$. The standard remarks about upper and lower central series all generalize in predictable fashion. Before proceeding, I must emphasize the connectivity of F. Where F is not mentioned explicitly, it corresponds to the surjectivity of $\pi_1(E) \to \pi_1(B)$ in the homotopy exact sequence of the fibration.

This class of fibrations draws its significance from the following theorem [21]

(4.4) *Let*

$$\begin{array}{ccc} F_1 & \xrightarrow{Ff} & F_2 \\ \downarrow & & \downarrow \\ E_1 & \xrightarrow{f} & E_2 \\ \downarrow{\scriptstyle p_1} & & \downarrow{\scriptstyle p_2} \\ B_1 & \xrightarrow{g} & B_2 \end{array}$$

be a map between quasi-nilpotent fibrations such that $\pi_1(g) : \pi_1(B_1) \to \pi_1(B_2)$ *is an isomorphism. If any two of* $H_*(Ff)$, $H_*(f)$, $H_*(g)$ *is an isomorphism, then so is the third.*

An important class of quasi-nilpotent fibrations $p : E \to B$ comprises those which are *nilpotent* [6], [20]. (Their usual definition, not needed here, is that $\pi_1(B)$ acts nilpotently on each of the homotopy groups of F_p.) These have Moore–Postnikov system admitting a refinement composed of principal fibrations – which means that p factors as the composition of a homotopy equivalence with fibrations $D_i \to D_{i-1}$ ($D_0 = B$) induced from path-fibrations $PK(A_i, n_i) \to K(A_i, n_i)$ where A_i is an abelian group and $n_i \geqslant 2$.

$$\begin{array}{ccccc} & & \vdots & & \\ & \nearrow^{\sim} & \downarrow & & \\ & & D_i & \longrightarrow & PK(A_i, n_i) \\ & & \downarrow & \lrcorner & \downarrow \\ E & & D_{i-1} & \longrightarrow & K(A_i, n_i) \\ & & \downarrow & & \\ & \searrow_{p} & \vdots & & \\ & & \downarrow & & \\ & & B & & \end{array}$$

We remark that such properties are preserved by pull-backs [38 (9.2.5)].

Application of (4.4) to the fibrations $F_f \to X \to Y$ and $* \to Y \to Y$ yields the final necessary and sufficient condition for $f : X \to Y$ to be acyclic.

(4.3) (concluded): (xii) f *is quasi-nilpotent and* $H_*(f)$ *is an isomorphism.*

This is such a valuable characterization of acyclic maps that it makes it worthwhile pausing here to note some basic properties of nilpotent and quasi-nilpotent maps. Assume that all spaces in question, including fibres of maps, are connected. We say that a space X is *nilpotent* if the map $X \to *$ is. By pulling-back $f : X \to Y$ to $F_f \to *$ we thus have immediately

(4.5) *If* $f : X \to Y$ *is nilpotent, then so is its fibre* F_f.

(4.6) [**6** II 4.4] *If any two of* $f : X \to Y, g : Y \to Z$ *and* $g \circ f$ *are nilpotent, then so is the third.*

This result, which we state without proof, straightaway implies the following.

(4.7) COROLLARY: *If any two of* X, Y *and* $f : X \to Y$ *are nilpotent, then so is the third.*

Two further theorems are worth quoting from the literature.

(4.8) [**11** (7.2)] *If* X *is nilpotent, then so is the fibre of* $f : X \to Y$.

(4.9) [**19** (2.2)] $f : X \to Y$ *is nilpotent if and only if it is quasi-nilpotent and its fibre* F_f *is nilpotent.*

(4.10) COROLLARY: *Suppose* $f : X \to Y$ *is a quasi-nilpotent fibration. Then* X *is nilpotent if and only if both* F_f *and* Y *are.*

Proof. The nilpotence of X implies in turn that of F_f (by (4.8)), f (4.9), and then Y (4.7). Conversely, if F_f is nilpotent, then so are f (4.9) and (given Y nilpotent) X (4.7).

A final observation here is that if X is nilpotent, then its fundamental group is (wait for it . . .) nilpotent. This is proved by induction, with each fibration $D_i \to D_{i-1} \to K(A_i, 2)$ $(D_0 = *)$ giving rise to a central extension [**38** (7.3.13)] $A \rightarrowtail \pi_1(D_i) \twoheadrightarrow \pi_1(D_{i-1})$ where $\pi_1(D_{i-1})$ is already nilpotent. Then a lower central series check forces $\pi_1(D_i)$ and so, eventually, $\pi_1(X)$ to be nilpotent. This implies that the derived series of $\pi_1(X)$ terminates at 1, whence

(4.11) *If* X *is nilpotent, then* $P\pi_1(X) = 1$.

(4.12) *Suppose* $f : X \to Y$ *is acyclic. Then* $g : Y \to Z$ *is acyclic if and only if* $g \circ f$ *is.*

This is immediate from (ii) of (4.3). Examples will be provided later on (4.19)(3), (5.12)(3) to show that g, $g \circ f$ acyclic do not imply f acyclic (in stark contrast to the situation for maps which merely induce integral homology isomorphisms). To draw such a conclusion we need to strengthen the assumptions, as follows.

(4.13) *Given* $f : X \to Y$, $g : Y \to Z$ *with* $H_*(f)$ *an isomorphism. Then* f *is acyclic provided either*

(a) $g, g \circ f$ *are nilpotent; or*

(b) $g, g \circ f$ *are quasi-nilpotent with the fibre* F_g *of* g *simply-connected.*

Proof. Both arguments are applications of (4.3) xii).

(a) If g and $g \circ f$ are nilpotent, then so is f (4.6).

(b) From (4.4) $Ff_* : H_*(F_{g \circ f}) \to H_*(F_g)$ is an isomorphism, so that, because $\pi_1(F_g) = 1$, Ff is acyclic. Then by (4.2) f is too.

The case (4.13) a) will be exploited more fully once we have witnessed the effect of acyclic maps on fundamental groups.

(4.14) *If* $f : X \to Y$ *is acyclic, then* $\pi_1(Y) \cong \pi_1(X)/P$ *where* P *is some perfect normal subgroup of* $\pi_1(X)$.

Proof. Let F be the homotopy fibre of f with $i : F \to X$. Then in the exact sequence

$$\pi_2(Y) \to \pi_1(F) \xrightarrow{\pi_1(i)} \pi_1(X) \xrightarrow{\pi_1(f)} \pi_1(Y) \longrightarrow \pi_0(F)\ ,$$

$\pi_1(f)$ is onto because $\widetilde{H}_0(F) = 0$ implies $\pi_0(F) = 1$ while Ker $\pi_1(f) = \text{Im}\ \pi_1(i)$. The fact that $H_1(F) = 0$ implies that $\pi_1(F)$ is perfect, and so its homomorphic image P is also perfect after (1.4) a).

In the case where P is trivial, the homomorphism $\pi_2(Y) \to \pi_1(F)$ maps an abelian group onto a perfect group, which must therefore be trivial. This makes F both acyclic and simply-connected, in other words contractible. As an exercise, the reader may also use (4.3) iv) to obtain our conclusion, namely

(4.15) *If* $f : X \to Y$ *is acyclic and* $\pi_1(f)$ *is an isomorphism, then* f *is a homotopy equivalence.*

This argument combines with (4.3)xii) to give a new generalized Whitehead theorem. (The aim of the game is to find weak assumptions on an homology equivalence – that is, a map inducing integral homology isomorphisms – which guarantee that it is a homotopy equivalence.)

(4.16) *If* $f : X \to Y$ *is a quasi-nilpotent homology equivalence and* $P\pi_1(X) = 1$, *then* f *is a homotopy equivalence.*

Now return to $f : X \to Y$ as in (4.13) a). Because f is acyclic, so too is $Ff : F_{g \circ f} \to F_g$ (4.2). However $F_{g \circ f}$ is nilpotent (4.5); thus $P\pi_1(F_{g \circ f}) = 1$ (4.11). So (4.16) applies to Ff, forcing it, and thence also f, to be a homotopy equivalence.

(4.17) *The conditions of* (4.13) a) *ensure that* f *is a homotopy equivalence.*

Specialization to the case Z = * yields the known result that an homology equivalence between nilpotent spaces is a homotopy equivalence. On the other hand, the case Z = Y and $g = id_Y$ yields the more general, new result that

(4.18) *A nilpotent homology equivalence is a homotopy equivalence.*

We can now summarize the key notions of this chapter, listing the most general first, as follows.

homology equivalence

acyclic map $\Longleftrightarrow$ quasi-nilpotent homology equivalence

homotopy equivalence $\Longleftrightarrow$ nilpotent homology equivalence

(4.14) and (4.15) give a pretty good idea of what examples of acyclic maps must look like. Ours are drawn first from geometry, then from algebra.

(4.19) EXAMPLES: (1) For $n \geqslant 2$, let M be a closed n-manifold which is an homology n-sphere. Because $H_n(M) \neq 0$, M is oriented, and so the map $f : M \to S^n$ defined by collapsing the complement of a neighbourhood U ($\cong \mathbb{R}^n$) of a small cell ($\cong B^n$) in M induces an isomorphism $H_*(f) : H_n(M) \to H_n(M, M\backslash U) \cong H_n(S^n)$. Whence f is acyclic, S^n being simply-connected.

(2) The Poincaré 3-sphere is derived from the faithful smooth representation of the perfect alternating group A_5 in SO(3), *i.e.* in (1) above let n = 3, $M = SO(3)/A_5 = S^3/SL(2,5)$ (which, it follows from [35], is the only example of (1) where M is covered by a sphere).

(3) For M as in (2), let ω be a non-trivial element of $\pi_1(M)$ (= binary icosahedral group) so that its action on $\pi_2(M \vee S^2)$ is non-trivial, *i.e.* if $j : S^2 \to M \vee S^2$ generates $\pi_2(S^2)$ as a direct

summand [38 pp 419, 420] of

$$\pi_2(M_\vee S^2) = \pi_2(S^2) \oplus \pi_3(M\times S^2, M_\vee S^2),$$

then $h_{[\omega]}(j) \neq j$. (From the universal covering space $S^3{}_\vee({}_\vee S^2)$, one sees the latter summand is $\mathbb{Z}^{119}$.) Now let $k : S^2 \to M_\vee S^2$ be a map whose homotopy class is $2j - h_{[\omega]}(j) \in \pi_2(M_\vee S^2)$; it therefore represents j in homology. So

$$H_*(k) : \widetilde{H}_*(S^2) \to \widetilde{H}_*(S^2) \oplus \widetilde{H}_*(M) = \widetilde{H}_*(M_\vee S^2)$$

is the inclusion of the former summand. Next form the mapping cone C_k (the adjunction space $B^3 \cup_k (M_\vee S^2)$). From the exact sequence for k

$$\ldots \to H_q(S^2) \xrightarrow{H_*(k)} H_q(M_\vee S^2) \longrightarrow H_q(C_k) \longrightarrow H_{q-1}(S^2) \to \ldots,$$

the inclusion $i : M \hookrightarrow B^3 \cup_k (M_\vee S^2)$ is an homology isomorphism. Moreover $f : M \to S^3$ defined in (1) extends over $B^3 \cup_k (M_\vee S^2)$ to $\overline{f}$ by collapsing the now enlarged complement of $U \subset M \subset B^3 \cup_k (M_\vee S^2)$. So $H_*(\overline{f}) = H_*(f) \circ (H_*(i))^{-1}$ is an isomorphism. This makes both $\overline{f}$ and $f = \overline{f} \circ i$ acyclic.

$$M \overset{i}{\hookrightarrow} B^3 \cup_k (M_\vee S^2) \xrightarrow{\overline{f}} S^3$$

However i cannot be acyclic, for $\pi_1(i)$ is an isomorphism while $\pi_2(i)$ is not, threatening a contradiction of (4.15).

(4) Consider the map $B\pi$ induced on classifying spaces by $\pi : GLUT \to GL(A\oplus A)$ as in (3.9). From (3.10) it is an homology equivalence. However on fundamental groups it yields the original short exact sequence of groups being classified, namely

$$MA \rightarrowtail GLUT \overset{\pi}{\twoheadrightarrow} GL(A\oplus A).$$

In other words, Ker $\pi_1(B\pi)$ = Ker π is the abelian group MA, which is most definitely not perfect. Thus by (4.14) $B\pi$ cannot be acyclic, and hence by (4.3) xii) is not quasi-nilpotent. We shall see in the next chapter how this gives rise to a further example, of an acyclic map q with domain $BGL(A\oplus A)$ such that $q \circ B\pi$ is also acyclic although $B\pi$ fails to be.

It is also highly profitable to discuss the cofibration analogue of (4.1).

(4.20) *Suppose* f_1 *is a cofibration in the co-Cartesian square*

$$\begin{array}{ccc} X & \xrightarrow{f_1} & Y_1 \\ {\scriptstyle f_0}\downarrow & \ulcorner. & \downarrow{\scriptstyle f_0'} \\ Y_0 & \xrightarrow[f_1']{} & Y_0 \cup_X Y_1 \end{array} \quad .$$

If f_i *is acyclic, then so is* f_i' $(i = 0, 1)$.

Proof. Let $\{L\}$ be a local coefficient system on $Y_0 \cup_X Y_1$. Because f_1 is a cofibration the right-hand vertical homomorphism in

$$\begin{array}{ccccc} \ldots \to H_q(X; f_1^* f_0'^*\{L\}) & \to & H_q(Y_1; f_0'^*\{L\}) & \to & H_q(Y_1, f_1(X); f_0'^*\{L\}) \to \ldots \\ \downarrow & & \downarrow & & \downarrow \\ \ldots \to H_q(Y_0; f_1'^*\{L\}) & \to & H_q(Y_0 \cup_X Y_1; \{L\}) & \to & H_q(Y_0 \cup_X Y_1, f_1'(Y_0); \{L\}) \to \ldots \end{array}$$

is an excision isomorphism. The results follow by chasing the above diagram and using characterization (4.3) ii).

Further information is supplied by the van Kampen theorem, to the effect that $\pi_1(Y_0 \cup_X Y_1) = \pi_1(Y_0) *_{\pi_1(X)} \pi_1(Y_1)$. So if f_1 is acyclic then $\pi_1(Y_1) = \pi_1(X)/\mathrm{Ker}\, \pi_1(f_1)$ (by (4.14)), whence $\mathrm{Ker}\, \pi_1(f_1')$ is the normal closure of $f_{0*}\, \mathrm{Ker}\, \pi_1(f_1)$.

(4.21) *If in* (4.20) $f_1 : X_1 \to Y_1$ *is an acyclic cofibration, then* $f_1' : Y_0 \to Y_0 \cup_X Y_1$ *is an acyclic cofibration with* $\mathrm{Ker}\, \pi_1(f_1')$ *the normal closure of the perfect subgroup* $f_{0*}\, \mathrm{Ker}\, \pi_1(f_1)$ *of* $\pi_1(Y_0)$.

5 The plus-construction

Now for the crucial classification theorem for acyclic maps from a space X (where X is as before). This will enable us to speak of "the" space obtained by an acyclic map killing a given perfect normal subgroup of $\pi_1(X)$. Actually, it pays to be choosey here. For although there is a nice enough classification theorem at this level of generality, its applications disappoint, in that diagrams tend to commute only up to homotopy. So instead we restrict attention to *acyclic cofibrations.* (It is reassuring to note that if $f : X \to Y$ is any acyclic map, then the inclusion of X in the mapping cylinder Z_f ($\simeq Y$) is an acyclic cofibration.) The effect is to operate in the category *under* X (objects cofibrations $g : X \to Z$, morphisms strictly commuting triangles

$$\begin{array}{ccc} & X & \\ {}^{g_1}\swarrow & & \searrow{}^{g_2} \\ Z_1 & \xrightarrow[h]{} & Z_2 \end{array} \quad),$$

which has the pleasant – and here, crucial – property that if $h : Z_1 \to Z_2$ under X is a homotopy equivalence, then it is a homotopy equivalence under X, that is, with all triangles apex X strictly commuting [4]. The upshot is that judicious use of cofibrations makes many diagrams commute which might otherwise have done so only up to homotopy.

(5.1) THEOREM: *Let* P *be a perfect normal subgroup of* $\pi_1(X)$. *Then there exists an acyclic cofibration* $f : X \to Y$ *with* $\operatorname{Ker} \pi_1(f) = P$. *If* $f' : X \to Y'$ *is another acyclic cofibration with* $\operatorname{Ker} \pi_1(f') = P$, *then there exists a homotopy equivalence* $g : Y \to Y'$ *such that* $g \circ f = f'$.

Proof. Existence. We first treat the case where $P = \pi_1(X)$. For each generator α_λ of $\pi_1(X)$ attach a 2-cell to X with characteristic map some $a_\lambda : S^1 \to X$ whose homotopy class is α_λ. (Axiom of choice.) (Or, more frugally, do this only for each member of a subset which *normally* generates $\pi_1(X)$.) The resulting space W is simply-connected (van Kampen) so that there is a Hurewicz isomorphism $\pi_2(W) \to H_2(W)$.

$$\begin{array}{ccc} \underset{\lambda}{\vee} S^1 & \longrightarrow & X \\ \downarrow & \ulcorner. & \downarrow \\ \underset{\lambda}{\vee} B^2 & \longrightarrow & W \end{array}$$

$$\ldots \to H_2(W) \to H_2(W, X) \to H_1(X) \to \ldots$$

Because X has perfect fundamental group $H_1(X) = 0$, while by excision $H_2(W,X) = H_2(\underset{\lambda}{\vee} B^2, \underset{\lambda}{\vee} S^1)$ $= \underset{\lambda}{\oplus} \mathbb{Z}$. Maps $b_\lambda : S^2 \to W$, one for each λ, may now be chosen so that their wedge makes the composition

$$\widetilde{H}_q(\vee S^2) \to \widetilde{H}_q(W) \to H_q(W, X)$$

an homology isomorphism (being, for $q = 2$, an epimorphism between free abelian groups on the same number of generators). So attach 3-cells to W by $\vee b_\lambda : \underset{\lambda}{\vee} S^2 \to W$ to form another simply-connected space Y.

$$\begin{array}{ccc} \underset{\lambda}{\vee} S^2 & \longrightarrow & W \\ \downarrow & \ulcorner. & \downarrow \\ \underset{\lambda}{\vee} B^3 & \longrightarrow & Y \end{array}$$

Because Y is simply-connected it has the fundamental group we are seeking and also, in order to establish that $X \hookrightarrow Y$ is acyclic, we need only check that $H_*(Y, X) = 0$ (by (403)(ii), (iii), (iv) or (xii)). By the 5-lemma (and excision) the induced map of exact sequences of triples $(\vee B^3, \vee S^2, \text{pt.})$ and (Y, W, X)

$$\begin{array}{ccccccc} \ldots \to H_q(\vee S^2, \text{pt.}) & \to & H_q(\vee B^3, \text{pt.}) & \to & H_q(\vee B^3, \vee S^2) & \to & \ldots \\ \downarrow{\scriptstyle H_q(\vee b_n)} & & \downarrow & & \downarrow & & \\ \ldots \to H_q(W, X) & \longrightarrow & H_q(Y, X) & \longrightarrow & H_q(Y, W) & \longrightarrow & \ldots \end{array}$$

is an isomorphism throughout. In particular $H_*(Y, X) = 0$ as required.

For the general case, let $X' \to X$ be a covering of X with $\pi_1(X') = P$. By the above there is an acyclic cofibration $f' : X' \to Y'$ where Y' is simply-connected. Form the cofibration $f : X \to Y$ by the push-out

$$\begin{array}{ccc} X' & \xrightarrow{f'} & Y' \\ \downarrow & \ulcorner. & \downarrow \\ X & \xrightarrow[f]{} & Y \end{array} \qquad ;$$

(4.21) shows that f is acyclic with Ker $\pi_1(f) = P$.

Uniqueness. An immediate consequence of the following useful lemma.

(5.2) *Suppose* $f : X \to Y$ *and* $g : X \to Z$ *are maps with f an acyclic cofibration and* Ker $\pi_1(f) \leqslant$ Ker $\pi_1(g)$. *Then there exists a map* $h : Y \to Z$ *such that* $h \circ f = g$; *moreover any two such are homotopic.*

Proof. It may be checked that g may be replaced by the inclusion of X in the associated mapping cylinder: the substitute has the virtue of being a cofibration. This granted, assume that g is itself a cofibration. Obviously we shall use (4.21).

$$\begin{array}{ccc} X & \xrightarrow{f} & Y \\ {\scriptstyle g}\downarrow & \ulcorner. & \downarrow{\scriptstyle g'} \\ Z & \xrightarrow[f']{} & Z \cup_X Y \end{array}$$

This implies that f′ is an acyclic cofibration in the above co-Cartesian square, and that Ker $\pi_1(f')$, the normal closure of $\pi_1(g)(\text{Ker}\ \pi_1(f))$, is trivial. So, by (4.15), f′ is a homotopy equivalence. Since the diagram commutes and both g and $g' \circ f$ are cofibrations, it follows that f′ is a homotopy equivalence under X (the whole point of our refinement to acyclic cofibrations). By composing the unique homotopy inverse under X to f′ with g′, we obtain the desired homotopy class under X represented by h. Moreover, because the universal property of $Z \cup_X Y$ forces any such map h to assume this form, the class of h must be unique.

As an application we use (4.21) and the fact that the inclusion of an acyclic space A in its cone CA is an acyclic cofibration.

(5.3) COROLLARY: *If* A *is an acyclic space, then any map* $d : A \to X$ *induces as its mapping cone (= homotopy cofibre) inclusion* $i : X \to C_d$ *an acyclic cofibration killing the normal closure of* Im $\pi_1(d)$ *in* $\pi_1(X)$.

For an alternative proof, let $f : X \to Y$ be the acyclic cofibration with $\text{Ker}\ \pi_1(f) = \text{Im}\ \pi_1(d)$ whose existence is guaranteed by (5.1). Then by (5.2) both f and i enjoy the same universal property with respect to maps $g : X \to Z$ such that $g \circ d$ is nulhomotopic.

If P is an arbitrary perfect normal subgroup of the fundamental group of X, then the space Y of (5.1) may be written as X_P^+, the plus-construction with respect to P. (Thus in (5.3) C_d is (up to homotopy equivalence under X) X_P^+ where P is the normal closure in $\pi_1(X)$ of $\text{Im}\ \pi_1(d)$.) The reader should be warned of forthcoming lapses of pedantry: one often refers to X_P^+ as enjoying some property (e.g. (5.11)) when one really means that at least one representative for such a space has the property. It need not follow that all representatives do.

Examples of the geometric applicability of the construction $X \to X_P^+$ occur in, say, [17]. For our purposes, however, it is enough to consider only the case where P is the *maximal* perfect subgroup $P\pi_1(X)$ of $\pi_1(X)$ – which from (1.4) b) we know must be normal. This particular case earns its own notation: the reference to P is dropped and the map $X \to X^+$ is called q_X. (It is for the reader to decide whom the letter q commemorates.) q_X may be described, after (5.1), as the terminal object, up to homotopy, in the category of acyclic cofibrations under X. The following properties are worth checking. (The first uses (4.14) and (1.7).)

(5.4) *Acyclic* $f : X \to Y$ *is equivalent to* q_X *if and only if* $P\pi_1(Y) = 1$.

Of course, the term "equivalent" in (5.4) has two interpretations, according as f is a cofibration or not.

(5.5) *Given* $f : X \to Y$, *there is a unique homotopy class of maps* $f^+ : X^+ \to Y^+$ *making the following square commute.*

$$\begin{array}{ccc} X & \xrightarrow{f} & Y \\ \downarrow q_X & & \downarrow q_Y \\ X^+ & \xrightarrow{f^+} & Y^+ \end{array}$$

This result is a consequence of (5.2) and (1.4) d). From (1.6) a),

$$P\pi_1(X \times Z) = P(\pi_1(X) \times \pi_1(Z)) = P\pi_1(X) \times P\pi_1(Z),$$

so that . . .

(5.6) $$(X \times Z)^+ = X^+ \times Z^+ \quad \textit{with} \quad q_{X \times Z} = (q_X, q_Z) .$$

Next, (5.5) and (5.6) ($Z = I$) combine to yield . . .

(5.7) *If* $f_0 \simeq f_1 : X \to Y$, *then* $f_0^+ \simeq f_1^+ : X^+ \to Y^+$.

Algebraists aware of an absence of algebra hereabouts may find some consolation in the proof of the next result, on the one-point union of spaces.

(5.8) $$(X_1 \vee X_2)^+ = X_1^+ \vee X_2^+ \; \textit{with} \; q_{X_1 \vee X_2} = q_{X_1} \vee q_{X_2} .$$

Proof. The interesting thing about the proof is its use of some relatively deep group theory. (Or have I overlooked a more elementary argument?) For, to show that $P\pi_1(X_1^+ \vee X_2^+) = 1$, we appeal to the theorem of Kurosch [15] to the effect that any subgroup, H say, of a free product (such as $\pi_1(X_1^+ \vee X_2^+) = \pi_1(X_1^+) * \pi_1(X_2^+)$) is itself a free product of a free group, F say, with conjugates (say L_1, L_2) of subgroups of the original free factors. Now the Mayer-Vietoris sequence for $BH = BF \vee BL_1 \vee BL_2$ reveals that

$$H_1(BH) \cong H_1(BF) \oplus H_1(BL_1) \oplus H_1(BL_2),$$

whence H perfect (that is, $H_1(BH) = 0$) forces F, L_1, L_2 to be perfect as well. However F, being free, must be trivial, while in our case L_1, L_2 are perfect only so long as their conjugates in $\pi_1(X_1^+), \pi_1(X_2^+)$ are, making them trivial too (5.4). Hence $\pi_1(X_1^+ \vee X_2^+)$ has only trivial perfect subgroups. (More generally of course, we have shown that $PG_1 = 1$ and $PG_2 = 1$ imply that $P(G_1 * G_2) = 1$.)

The proof that $q_{X_1} \vee q_{X_2}$ is acyclic is more routine. Letting $\sqcup$ denote *disjoint* union, apply (4.21) to the co-Cartesian square

$$\begin{array}{ccc} X_1 \sqcup X_2 & \xrightarrow{q_{X_1} \sqcup q_{X_2}} & X_1^+ \sqcup X_2^+ \\ \downarrow & \ulcorner . & \downarrow \\ X_1 \vee X_2 & \longrightarrow & X_1^+ \vee X_2^+ \end{array} .$$

Note the fundamental group epimorphism

$$\pi_1(X_1) * \pi_1(X_2) \twoheadrightarrow \pi_1(X_1) \times \pi_1(X_2)$$

induced from the inclusion $j : X_1 \vee X_2 \hookrightarrow X_1 \times X_2$. Because the *smash product* $X_1 \wedge X_2 = (X_1 \times X_2)/(X_1 \vee X_2)$ is homotopy equivalent to the cofibre C_j of this inclusion [38 p.418], the van Kampen theorem for the push-out

$$\begin{array}{ccc} X_1 \vee X_2 & \longrightarrow & C(X_1 \vee X_2) \\ {\scriptstyle j}\downarrow & \ulcorner & \downarrow \\ X_1 \times X_2 & \longrightarrow & C_j \end{array}$$

yields that $\pi_1(X_1 \wedge X_2)$ is trivial. Meanwhile, it follows from the 5-lemma on the homology exact sequences of the pairs $(X_1 \times X_2, X_1 \vee X_2)$, $(X_1^+ \times X_2^+, X_1^+ \vee X_2^+)$ that $q_{X_1} \wedge q_{X_2} : X_1 \wedge X_2 \to X_1^+ \wedge X_2^+$ is an homology equivalence. Hence . . .

(5.9) $$(X_1 \wedge X_2)^+ = X_1 \wedge X_2 \simeq X_1^+ \wedge X_2^+ .$$

(5.10) *Suppose* $f : X \to Y$ *has path-connected homotopy fibre* F_f. *Then*

(a) *so has* f^+; *and*

(b) $P\pi_1(F_{f^+}) = 1$.

Proof. (a) Clearly X connected implies X^+ connected (consider zero-th homology). So chase the diagram

$$\begin{array}{ccccc} \pi_1(Y) & \to & \pi_0(F_f) & \to & \pi_0(X) \\ \downarrow & & \downarrow & & \downarrow \\ \pi_1(Y^+) & \to & \pi_0(F_{f^+}) & \to & \pi_0(X^+) . \end{array}$$

(b) Observe that $P\pi_1(X^+) = 1$ while $\pi_2(Y^+)$ is abelian in the exact sequence

$$\pi_2(Y^+) \to \pi_1(F_{f^+}) \to \pi_1(X^+) .$$

We shall shortly have much more to say about the effect of the plus-construction functor on a fibration. For the moment though it is more interesting to see how it behaves with respect to pushing-out. The following application of (4.21) highlights the relevance of perfect-radical-preserving epimorphisms, studied in (1.6).

(5.11) *Let* $f : X \to Y$ *have path-connected homotopy fibre. Then the square*

$$\begin{array}{ccc} X & \xrightarrow{q_X} & X^+ \\ f \downarrow & & \downarrow f^+ \\ Y & \xrightarrow[q_Y]{} & Y^+ \end{array}$$

is co-Cartesian if and only if the epimorphism $\pi_1(f)$ *preserves perfect radicals.*

If however $\pi_1(f)$ is not known to be surjective, then (4.21) stipulates the appropriate necessary and sufficient condition as the normal closure of $\pi_1(f)(P\pi_1(X))$ being $P\pi_1(Y)$. This condition reappears in Chapter 9.

(5.12) EXAMPLES: (1) X^+ is contractible if and only if X is acyclic.

(2) In (4.19)(1), $f : M \to S^n$ is just $q_M : M \to M^+$.

(3) This continues Example 4.19(4). In the commuting square

$$\begin{array}{ccc} BGLUT & \xrightarrow{q_1} & BGLUT^+ \\ B\pi \downarrow & & \downarrow B\pi^+ \\ BGL(A\oplus A) & \xrightarrow[q_2]{} & BGL(A\oplus A)^+ \end{array}$$

(where $q_1 = q_{BGLUT}$, etc.), $B\pi^+$ is an homology equivalence, between spaces which will be shown in (11.6) to be nilpotent. Therefore (4.18) $B\pi^+$ is a *homotopy equivalence*, making $B\pi^+ \circ q_1$ a composition of acyclic maps and thus, by (4.12), acyclic. This provides the example promised in (4.19)(4), for q_2 and $q_2 \circ B\pi$ are acyclic although $B\pi$ is not.

6 The plus-construction on a fibration

Suppose we have a space E whose homotopy groups are only accessible through the homotopy exact sequence of a fibration $F \to E \to B$. Chances are, we're going to know more about the homotopy groups of F^+ and B^+ than those of E^+. For them to be of much use however we need to relate them nicely to $\pi_*(E^+)$. So what we'd really like is to be able to say that $F \to E \to B$ is a fibration implies $F^+ \to E^+ \to B^+$ is also a fibration. The justification of such statements is the goal of this chapter.

First, the case of non-connected fibres. Here (and elsewhere) we need to recall some well-known homotopy theory concerning the classification of (connected) covering spaces. On the one hand, there are strong uniqueness properties (see, *e.g.*, [38 (2.5.3)]), which enable one to speak of "the" covering of X associated to any given normal subgroup of $\pi_1(X)$. On the other, by the adjunction of (finitely many) 3-cells to the space X, one can kill off all elements of $\pi_2(X)$; then addition of (finitely many) 4-cells annihilates π_3 of the new space without altering π_1, π_2. Iteration of this procedure gives rise to another space in our category which has trivial homotopy groups, apart from the same fundamental group as X. Thus it serves as a model for the Eilenberg-Maclane space $K(\pi_1(X), 1)$, and comes complete with a map $X \to K(\pi_1(X), 1)$ (inclusion, in fact) which induces the identity homomorphism on fundamental groups. The usual model for a space K(G, 1) is Milnor's classifying space BG, obtained by factoring out the (contractible) infinite join $EG = G*G*\dots$ by diagonal G-multiplication. The homotopy groups of BG may be checked from the fact that EG forms a covering space over BG (G being discrete), and so, by uniqueness, *the* universal cover of BG. Note that any two models for K(G, 1) have the same homotopy type (Whitehead theorem) – prompting us to cheat occasionally and let $B1 = pt$ –, while $G \mapsto BG$ forms a functor (inverse to $X \mapsto \pi_1(X)$). In Chapter 11 we shall use the further property that an extension $N \rightarrowtail G \twoheadrightarrow Q$ induces a *fibre sequence* $BN \to BG \to BQ$ (*i.e.* BN has the homotopy type of the homotopy fibre of $BG \to BQ$), as a check of homotopy groups reveals.) In sum then, if $H \trianglelefteq \pi_1(X)$, then there is a map $X \to K(\pi_1(X)/H, 1) = K$ (factoring through $X \to K(\pi_1(X), 1)$) whose homotopy fibre $X \times_K PK$ is the cover of X with fundamental group equal (from the exact sequence of the fibration) to H.

(6.1) PROPOSITION: *Let* $\bar{X}$ *be the covering space of* X *corresponding to some normal subgroup* H *of* $\pi_1(X)$ *containing* $P\pi_1(X)$. *Then* $\bar{X}^+$ *is the covering of* X^+ *corresponding to the normal subgroup* $H/P\pi_1(X)$ *of* $\pi_1(X)/P\pi_1(X) = \pi_1(X^+)$.

Proof. By (5.2), the classifying map $X \to K(\pi_1(X)/H, 1)$ for the covering $\bar{X}$ factors through X^+, where the induced covering is Y, say.

$$\begin{array}{ccccc} \bar{X} & \xrightarrow{\bar{q}} & Y & \longrightarrow & PK(\pi_1(X)/H, 1) \\ \downarrow & & \downarrow & & \downarrow \\ X & \xrightarrow[q_X]{} & X^+ & \longrightarrow & K(\pi_1(X)/H, 1) \end{array}$$

Then $\bar{q} : \bar{X} \to Y$ is acyclic after (4.1), while Ker $\pi_1(\bar{q})$ = Ker $\pi_1(q_X) = P\pi_1(X) = PH$; so finish via (5.1).

(6.2) COROLLARY: *Let* $p : E \to B$ *have path-connected homotopy fibre* F. *Let* $k : \bar{B} \to B$ *be the covering corresponding to* $P\pi_1(B)$, *with pull-back* $\bar{p} : \bar{E} \to \bar{B}$ *of* p *over* k. *Then* $F_{p^+} = F_{\bar{p}^+}$.

Proof. We have commuting squares

$$\begin{array}{ccc} \bar{E} & \longrightarrow & E \\ \downarrow{\scriptstyle \bar{p}} & & \downarrow{\scriptstyle p} \\ \bar{B} & \xrightarrow{k} & B \end{array} \qquad\qquad \begin{array}{ccc} \bar{E}^+ & \longrightarrow & E^+ \\ \downarrow{\scriptstyle \bar{p}^+} & & \downarrow{\scriptstyle p^+} \\ \bar{B}^+ & \xrightarrow{k^+} & B^+ \end{array}$$

of which the former is known to be a pull-back; we prove the latter is too. A diagram chase of the homotopy exact sequences

$$\begin{array}{ccccc} \pi_1(F) & \longrightarrow & \pi_1(\bar{E}) & \twoheadrightarrow & \pi_1(\bar{B}) = P\pi_1(B) \\ \| & & \downarrow & & \downarrow \\ \pi_1(F) & \longrightarrow & \pi_1(E) & \longrightarrow & \pi_1(B) \end{array}$$

shows that $\bar{E}$ is the covering of E associated to some normal subgroup of $\pi_1(E)$ which contains $P\pi_1(E)$. Then by (6.1) both horizontal maps in the right-hand square above are covering projections with the same fibre (because $\pi_1(E)/\pi_1(\bar{E}) \cong \pi_1(B)/\pi_1(\bar{B})$). Uniqueness of covering projections now obliges $\bar{E}^+$ to be $\bar{B}^+ \times_{B^+} E^+$. Thus $\bar{p}^+$ is indeed a pull-back of p^+ and so has the same homotopy fibre.

Since it is the homotopy groups of X^+ that we're after, we may allow ourselves a little freedom in choice of the space which is to represent the homotopy type of X^+. In particular this enables us to replace maps by their associated mapping-path fibrations if desired, and to speak simply of fibres rather than homotopy fibres. Also, this means we refer to a "fibration" at times when a purist would use the term "fibre sequence".

A useful device for the main result on the plus-construction and fibrations, (6.4) below, is that of the *fibrewise +-construction.* The next lemma sets this up.

(6.3) LEMMA: *If* $F \xrightarrow{i} E \xrightarrow{p} B$ *is a fibration, then so is* $F^+ \to E \cup_F F^+ \to B$.

Proof. After (5.2), the trivial map $F \to B$ factors through F^+. Then the universal property of push-outs enables us to define $p_1 : E \cup_F F^+ \to B$, whose fibre we seek to prove is F^+.

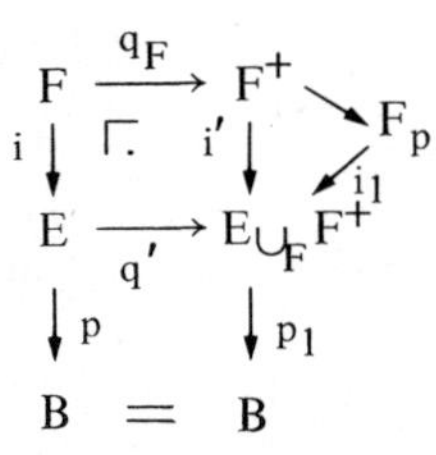

Now (4.20) and (4.2) guarantee that $q' : E \to E \cup_F F^+$ and in turn $Fq' : F \to F_{p_1}$ are acyclic. If we knew that Fq' factored through F^+ then (5.1) would immediately finish the proof. As we don't, the argument is less direct; it shows that $P\pi_1(F_{p_1}) = 1$, or equally (from (1.6), Ker $\pi_1(i_1)$ being central in $\pi_1(F_{p_1})$ [38 (7.3.13)]), that $i_{1*}P\pi_1(F_{p_1}) = 1$ in $\pi_1(E \cup_F F^+)$. Because $\pi_1(Fq')$ is an epimorphism with perfect kernel (4.14), $P\pi_1(F_{p_1}) = (Fq')_* P\pi_1(F)$ (using (1.6) again), so that

$$i_{1*}P\pi_1(F_{p_1}) = i_{1*}(Fq')_*P\pi_1(F) = q'_* i_* P\pi_1(F) = i'_* q_{F*} P\pi_1(F) = 1,$$

as $P\pi_1(F)$ is precisely what $\pi_1(q_F)$ kills.

Now for the key result.

(6.4) THEOREM: *If* $F \xrightarrow{i} E \xrightarrow{p} B$ *is a fibration (of connected spaces), then so is* $F^+ \xrightarrow{i^+} E^+ \xrightarrow{p^+} B^+$, *provided either*

(a) $P\pi_1(B) = 1$ *or*

(b) p *is quasi-nilpotent and* F^+ *is nilpotent.*

Proof. (a) is easy (but worthwhile). For here $B^+ = B$, so that we are considering the map $q_E : E \to E^+$ over B. After (4.2) the map of fibres $Fq_E : F \to F_{p^+}$ is acyclic. However, by (5.10), $P\pi_1(F_{p^+}) = 1$. So by (5.4) F_{p^+} is F^+ after all.

(b) Note that p nilpotent is a special case, by virtue of (4.9) which ensures that F is nilpotent and so (4.11) $F^+ = F$ is too. We postpone the proof of this case until after we have seen how it leads to the more general result. To see this therefore, apply (6.3) to obtain the fibration $F^+ \to E \cup_F F^+ \xrightarrow{p_1} B$. Certainly p_1 is quasi-nilpotent, because $\pi_1(B)$ is known to act nilpotently on $H_*(F) = H_*(F^+)$. But now that p_1 has nilpotent fibre it is further nilpotent (4.9). Given that this forces a fibration $(F^+)^+ \to (E \cup_F F^+)^+ \to B^+$, the conclusion follows from $(F^+)^+ = F^+$ and also (because $E \to E \cup_F F^+$ is acyclic) $(E \cup_F F^+)^+ = E^+$.

Now for the proof when p is nilpotent. We appeal to the lemma (6.5) below to get the nilpotence, in turn, of $p^+ : E^+ \to B^+$ and thence its pull-back $p' : B \times_{B^+} E^+ \to B$. The maps $p : E \to B$, $q_E : E \to E^+$ determine $s : E \to B \times_{B^+} E^+$ and so the following commuting diagram.

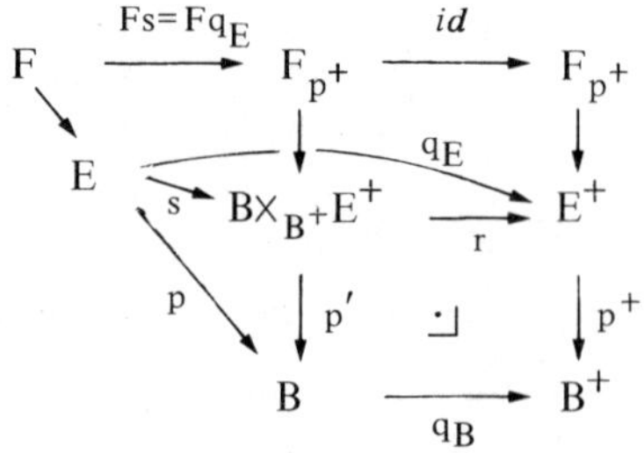

After (4.1) $r : BX_B E^+$ is acyclic. Thus $H_*(s)$ is an isomorphism because $H_*(r)$, $H_*(q_E)$ are. Then (4.17) shows that s is a homotopy equivalence, whence Fs is too. This completes the proof because F, being nilpotent, is already F^+.

(6.5) LEMMA: *If* $f : X \to Y$ *is a nilpotent fibration, then* $f^+ : X^+ \to Y^+$ *is also nilpotent.*

Proof. (5.10) a) ensures that F_{f^+} is path-connected. By our definition of a nilpotent fibration, the Moore-Postnikov system of f refines to principal fibrations. So we may restrict attention to this case, needing to prove only that if $X \to Y \to K(A, n)$, $n \geqslant 2$, is a fibration, then so is $X^+ \to Y^+ \to K(A, n)$. This is immediate from (6.4) a).

Note that the hypothesis (6.4) b) is stronger than need be. For by (6.2) in order to show that $F_{p^+} = F^+$ it suffices to establish it with p replaced by $\bar{p}$. The condition that p be nilpotent may therefore be weakened to the action of $P\pi_1(B)$ (rather than that of the whole of $\pi_1(B)$) on $H_*(F)$ being nilpotent. As a bonus, the weakened hypothesis admits a surprising simplification. . .

(6.6) *If a perfect group* π *acts nilpotently on a group* N, *then* π *acts trivially on* N *(and conversely).*

Proof. (We first dismiss the converse as trivial.) We show that π acts trivially on each $\zeta_i^\pi N$ (see p.37). Suppose done for $i = k \geqslant 0$. For any $g \in \zeta_{k+1}^\pi N$ the π-action defines a function ρ_g from π to $\zeta_k^\pi N$ by $\theta \mapsto [g, \theta]$. Because $[\pi, \zeta_k^\pi N] = 1$, we have

$$\begin{aligned}[g, \theta_1][g, \theta_2] &= g\theta_1 g^{-1}\theta_1^{-1}[g, \theta_2] \\ &= g\theta_1 g^{-1}[g, \theta_2]\theta_1^{-1} \\ &= [g, \theta_1\theta_2].\end{aligned}$$

This makes ρ_g a homomorphism. Since the only homomorphism from a perfect group to an abelian group is the trivial one, it follows that g commutes with π in $N \rtimes \pi$, so that the induction goes through.

(6.7) EXAMPLES: (1) Let $G = SL(2, 5)$. Application of (6.4) a) to the fibration

$$S^3/G \to BG \to BS^3$$

gives a description (after (5.12)(2) and (4.19)(2)) of $BSL(2, 5)^+$ as an S^3-fibre space over $BS^{3+} = BS^3$, the infinite-dimensional quaternionic projective space.

(2) Recall from (4.19)(4) that there is a fibration

$$BMA \to BGLUT \xrightarrow{B\pi} BGL(A \oplus A).$$

Now MA is abelian, so that $BMA = K(MA, 1) = \Omega K(MA, 2)$ is a nilpotent space (after, for example, (4.8) because it is the fibre of $PK(MA, 2) \to K(MA, 2)$ where $PK(MA, 2)$ is contractible, hence nilpotent). Were $B\pi$ quasi-nilpotent, which by (4.19)(4) it isn't, we would then deduce from (6.4) b) that there is a fibration

$$BMA \longrightarrow BGLUT^+ \longrightarrow BGL(A \oplus A)^+ ,$$

which by (5.12)(3) there isn't.

In (6.4) we obtained sufficient conditions for the plus-construction to preserve fibrations. The following necessary condition further explains our preoccupation in (1.6) with epimorphisms which preserve perfect radicals.

(6.8) *If both* $F \xrightarrow{i} E \xrightarrow{p} B$ *and* $F^+ \xrightarrow{i^+} E^+ \xrightarrow{p^+} B^+$ *are fibrations (of path-connected spaces), then*

$$p_* P\pi_1(E) = P\pi_1(B).$$

Proof. Certainly, after (1.4) c), $p_* P\pi_1(E) \leqslant P\pi_1(B)$. There is in consequence a commuting diagram with both columns and rows exact:

$$\begin{array}{ccccc}
 & & P\pi_1(E) & \xrightarrow{p_*|} & P\pi_1(B) \\
 & & \downarrow & & \downarrow \\
\pi_1(F) & \xrightarrow{i_*} & \pi_1(E) & \xrightarrow{p_*} & \pi_1(B) \\
\downarrow q_{F*} & & \downarrow q_{E*} & & \downarrow q_{B*} \\
\pi_1(F^+) & \xrightarrow{i^+_*} & \pi_1(E^+) & \xrightarrow{p^+_*} & \pi_1(B^+)
\end{array}$$

Now both p_* and q_{F*} are onto. So too therefore is $p_*|$, as a diagram chase reveals.

Note that hypothesis (a) of (6.4) trivially satisfies the condition on $\pi_1(p)$, while a nilpotent p satisfies it by virtue of (1.6) b) (*cf.* remarks preceding (11.10)). I have managed to refrain from discussing (6.4) b) in general in this light (!).

7 The acyclic space construction

In furtherance of our strategy of getting a grip on the homotopy groups of X^+ by identifying fibrations of which it is a part, we now determine the fibre of $q_X : X \to X^+$. Since q_X is functorial, so too must be the construction of its fibre. Following [12], we describe the acyclic fibre, $AX = X_\infty$ as an inverse limit of spaces X_n over X, each in turn killing off a bit more of the homology of X than those under it. (Note that our X_n = Dror's $A_{n-1}X$, the difference in numeration being a straightforward reflection of our interest in homotopy groups of X^+ rather than of $AX = A_\infty X$. Compare (8.1) below.)

This means we take $X_1 = X$, and X_2 as the covering space of X with $\pi_1(X_2) = P\pi_1(X)$. Then $X_3, X_4, \ldots$ are constructed as follows.

(7.1) PROPOSITION: *There is a sequence of spaces*

$$\ldots \to X_{n+1} \to X_n \to \ldots \to X_3 \to X_2$$

such that

(i) $\tilde{H}_q(X_n) = 0, q < n$;

(ii) $X_{n+1} \to X_n$ *is induced from the path fibration* $(n \geq 2)$

$$K(H_n(X_n), n-1) = \Omega K(H_n(X_n), n) \to PK(H_n(X_n), n) \to K(H_n(X_n), n)$$

(*that is,* $X_{n+1} = X \times_{K(H_n(X_n),n)} PK(H_n(X_n), n)$); *and*

(iii) X_n *is unique up to fibre homotopy equivalence (over* X_{n-1}*), and the construction is functorial up to fibre homotopy.*

Remark. It is easier to check what properties (iii) alludes to than to define the appropriate categories on which the construction behaves as a true functor should.

Proof. Certainly X_2 does what is required of it. We therefore consider $X_{n+1} \to X_n$, $n \geq 2$, supposing $X_n \to X_{n-1} \to \ldots \to X_2$ to have been constructed satisfying (i), (ii) and (iii). At our disposal is the choice of map $\theta_n : X_n \to K(H_n(X_n), n)$. Since, by hypothesis, $H_{n-1}(X_n) = 0$, we have from the universal coefficient sequence [**38** (5.5.3)] a natural isomorphism

$$\mathrm{Hom}(H_n(X_n), H_n(X_n)) \to H^n(X_n; H_n(X_n)).$$

But there is also a natural isomorphism [**38** (8.1.8)]

$$H^n(X_n; H_n(X_n)) \to [X_n, K(H_n(X_n), n)] \ .$$

The composition of these, applied to the identity homomorphism in $\mathrm{Hom}(H_n(X_n), H_n(X_n))$, gives a natural determination of θ_n up to homotopy. As (ii) dictates, $X_{n+1} \to X_n$ is defined to be the pull-back of the path-fibration over ϑ_n (in other words, X_{n+1} is the fibre of θ_n). Thus, variation of θ_n within its homotopy class causes variation of X_{n+1} within its fibre homotopy class.

Finally, the triviality of $H_q(X_{n+1})$, $q \leqslant n$, is established by a Serre spectral sequence argument. For example, $n \geqslant 2$ implies that $K(H_n(X_n), n)$ is simply-connected and consequently its path-fibration *orientable.* So $K = K(H_n(X_n), n-1) \to X_{n+1} \to X_n$ is also orientable; its Serre homology sequence

$$H_{2n-2}(K) \to \ldots \to H_n(K) \to H_n(X_{n+1}) \to H_n(X_n) \xrightarrow{\sim} H_{n-1}(K) \to \ldots$$

forces $\widetilde{H}_q(X_{n+1}) = 0$ for $q < n+1$ because, by the Hurewicz isomorphism theorem,

$$\widetilde{H}_q(K(H_n(X_n), n-1)) = \begin{cases} 0 & q < n-1, \\ H_n(X_n) & q = n-1, \\ 0 & q = n. \end{cases}$$

Here, $H_n(K) = 0$ follows by applying the Hurewicz theorem to the $(n-1)$-skeleton K^{n-1} of K, and the exact sequences of (K, K^{n-1}):

$$\begin{array}{ccccccc} O = \pi_n(K) & \longrightarrow & \pi_n(K, K^{n-1}) & \longrightarrow & \pi_{n-1}(K^{n-1}) \\ \downarrow & & \downarrow\wr & & \downarrow\wr \\ O = H_n(K^{n-1}) \longrightarrow H_n(K) & \longrightarrow & H_n(K, K^{n-1}) & \longrightarrow & H_{n-1}(K^{n-1}) \end{array}$$

Of course, (K, K^{n-1}) is $(n-1)$-connected [**38** (7.6.16)]. Note that, even were $\pi_n(K)$ non-trivial, we could still infer G. Whitehead's generalization of the Hurewicz theorem: $\pi_n(K) \twoheadrightarrow H_n(K)$.

(7.2) COROLLARY: *If* $\widetilde{H}_q(Z) = 0$ *whenever* $q \leqslant n$, *then* $X_{n+1} \to X$ *induces a bijection* $[Z, X_{n+1}] \to [Z, X]$.

Proof. (I am tempted to leave this as an exercise – but the result is too valuable.) Of course one argues by induction on n. For n = 1, Z has perfect fundamental group, implying that its image in $\pi_1(X)$ is also perfect, therefore contained in $P\pi_1(X) = \mathrm{Im}[\pi_1(X_2) \hookrightarrow \pi_1(X)]$. Hence any (pointed) map from Z to X satisfies the conditions of the lifting theorem for covering projections and so lifts uniquely to X_2. For $n > 1$, we note that, because $X_{n+1} = X_n \times_K PK$ (where $K = K(H_n(X_n), n)$), the obstruction to lifting a map $Z \to X_n$ to X_{n+1} lies in $[Z, K] = H^n(Z; H_n(X_n))$, while that to lifting a homotopy between the projections to X_n of two given (pointed) maps to X_{n+1} lies in

$$[(Z\times I, Z\times \dot{I} \cup z_0 \times I), K] = H^n((Z, z_0)\times(I, \dot{I}); H_n(X_n)) = H^{n-1}(Z; H_n(X_n)) .$$

(Here $z_0 \in Z$ is the long-suppressed basepoint struggling to assert itself.) The universal coefficient theorem [**38** (5.5.3)] shows both cohomology groups vanish. One may also argue from the exact sequence (of sets) [**38** p. 461]

$$(7.3) \qquad [Z, K(H_n(X_n), n-1)] \to [Z, X_{n+1}] \to [Z, X_n] \to [Z, K(H_n(X_n), n)] .$$

Here the extreme left and right terms are again isomorphic respectively to the trivial groups $H^{n-1}(Z; H_n(X_n))$ and $H^n(Z; H_n(X_n))$ [**38** (8.1.8)].

One important application is to smash products.

(7.4) COROLLARY: *A homotopy class of maps* $S \wedge T \to X$ *induces a unique homotopy class of maps* $S_m \wedge T_n \to X_{m+n}$ *such that*

$$\begin{array}{ccc} S_m \wedge T_n & \longrightarrow & X_{m+n} \\ \downarrow & & \downarrow \\ S \wedge T & \longrightarrow & X \end{array}$$

commutes.

Proof. We show that $H_q(S_m \wedge T_n) = 0$ for $0 < q < m+n$. The push-out construction for the one-point union $S_m \vee T_n$ gives rise to a Mayer-Vietoris sequence, from which

$H_q(S_m \vee T_n) = H_q(S_m) \oplus H_q(T_n)$. The Künneth formula decomposes $H_q(S_m \times T_n)$ into a direct sum whose summands, with four exceptions, have the form $H_i(S_m) \otimes H_{q-i}(T_n)$ (trivial since $0 < i < m$ or $0 < q-i < n$) or Tor $(H_i(S_m), T_{q-i-1}(T_n))$ (similarly, trivial). The four exceptions differ from the above by involving a zero-th homology group. Because such a group is torsion-free the two Tor terms vanish, leaving

$$H_q(S_m \times S_n) = H_q(S_m) \otimes \mathbb{Z} \oplus \mathbb{Z} \otimes H_q(T_n) \cong H_q(S_m \vee T_n) .$$

Then the result follows from the homology exact sequence for the pair $(S_m \times T_n, S_m \vee T_n)$.

We are now ready to climb to the top of the tower. From (7.3) with Z a sphere it is clear that X_{n+1} differs from X_n only in respect of $(n-1)$- and n-th homotopy groups. So to gain information about any particular homotopy (or homology) group of the inverse limit space

$$AX = \varprojlim X_n$$

it is only necessary to climb finitely many steps of the tower. From (7.1) (since each such step is a nilpotent fibration, a property preserved under finite composition (4.6)), and (7.2), we have . . .

(7.5) PROPOSITION: (i) AX *is an acyclic space*;

(ii) $AX \to X_2$ *is a nilpotent fibration*;

(iii) *for any acyclic space* Z, *the fibration* $d_X : AX \to X$ *induces an isomorphism*

$$[Z, AX] \to [Z, X] ;$$

and in particular

(iv) *for any map* $f : W \to X$, *any map* $f' : AW \to AX$, *such that*

$$\begin{array}{ccc} AW & \xrightarrow{f'} & AX \\ {\scriptstyle d_W}\downarrow & & \downarrow{\scriptstyle d_X} \\ W & \xrightarrow{f} & X \end{array}$$

commutes, is homotopic over X *to* $Af : AW \to AX$.

Here Af is determined (non-uniquely) by means of (7.1) iii).

If the covering space $p : \overline{X}_P \to X$ corresponds to a normal perfect subgroup P of $\pi_1(X)$, then $p \circ d_{\overline{X}_P} : A\overline{X}_P \to X$ enjoys, by (7.5) iii), a property dual to $X \to X_P^+$. Likewise, (5.3) dualizes.

(7.6) PROPOSITION: *Any acyclic map* $f : X \to Y$ *with* $\text{Ker}\, \pi_1(f) = P \leqslant \pi_1(X)$ *has* $p \circ d_{\overline{X}_P} : A\overline{X}_P \to X$ *as homotopy fibre (up to homotopy equivalence over* X).

Proof. The two fibrations $p \circ d_{\overline{X}_P} : A\overline{X}_P \to X$ and $F_f \to X$ share the same universal property with respect to maps $d : A \to X$ such that A is acyclic and $f \circ d$ is nulhomotopic. (That $f \circ (p \circ d_{\overline{X}_P})$ is nulhomotopic follows from (5.3), (5.1).)

In combination, (5.3) and (7.6) assert that $A\overline{X}_P \to X \to X_P^+$ is both a fibre sequence and a cofibre sequence. The case where P is maximal rates a special mention.

(7.7) THEOREM: $AX \xrightarrow{d_X} X \xrightarrow{q_X} X^+$ *is both a fibre sequence and a cofibre sequence.*

In the usual way [38 p.461], the fibre sequence extends to the left, with ΩX^+ therefore the homotopy fibre of d_X, a fact to be used repeatedly in Chapter 10. (Incidentally, we may safely write ΩX^+ for $\Omega(X^+)$, since $\pi_1(\Omega X) = \pi_2(X)$ abelian implies $(\Omega X)^+ = \Omega X$.)

Finally, we can apply (6.4) a) to obtain fibrations

$$X_2^+ \longrightarrow X^+ \longrightarrow K(\pi_1(X^+), 1),$$

and, for $n \geqslant 2$,

$$X_{n+1}^+ \longrightarrow X_n^+ \longrightarrow K(H_n(X_n), n).$$

By means of the resulting homotopy exact sequences and the generalized Hurewicz isomorphism theorem, the following may be deduced by induction.

(7.8) *For* $n \geqslant 2$,

$$\pi_q(X_n^+) = \begin{cases} 0 & q < n, \\ \pi_q(X^+) & q \geqslant n, \end{cases}$$

while

$$\pi_n(X_n^+) \xrightarrow{\sim} H_n(X_n) \; ,$$
$$\pi_{n+1}(X_n^+) \longrightarrow\!\!\!\!\!\rightarrow H_{n+1}(X_n) \; .$$

8 The plus-construction on a classifying space

We have nearly arrived at our study of the homotopy groups of $BGLA^+$. First though, we pause for a more general study of the classifying space BG and the homotopy groups of BG^+ when G is assumed only to be a discrete group. This requires the results of the previous chapter when $X = BG = K(G, 1)$ and thus $H_*(X) = H_*(G)$.

(8.1) THEOREM:

$$\pi_q(BG^+) = \begin{cases} G/PG & q = 1 , \\ H_q((BG)_q) & q \geqslant 2. \end{cases}$$

Proof. Although the theorem itself follows immediately from (7.8), we pursue another proof which is more revealing about the space X_3. The method involves calculation of the homotopy groups of ABG from (7.1) and then those of BG^+ from (7.7). First observe that $X_2 = BPG$ and the homotopy exact sequence of the fibration $X_3 \to BPG \to K(H_2(X_2), 2)$ shows that $\pi_q(X_3) = 0$ whenever $q \geqslant 2$ (that is, $X_3 = K(\pi_1(X_3), 1)$), while $H_2(X_2) = \text{Ker}\,(\pi_1(X_3) \to PG)$. We can argue by induction to deduce generally that for $n \geqslant 3$

$$\pi_q(X_n) = \begin{cases} \pi_1(X_3) & q = 1 , \\ H_{q+1}(X_{q+1}) & 2 \leqslant q \leqslant n-2, \\ 0 & q \geqslant n-1 , \end{cases}$$

from the exact sequence of the fibration $X_{n+1} \to X_n \to K(H_n(X_n), n)$, namely ($\delta$ the Kronecker delta)

$$\pi_{q+1}(X_n) \to \delta_{q+1,n} \cdot H_n(X_n) \to \pi_q(X_{n+1}) \to \pi_q(X_n) \to \delta_{q,n} \cdot H_n(X_n) \ .$$

Consequently,

$$\pi_q(A BG) = \begin{cases} \pi_1(X_3) & q = 1 , \\ H_{q+1}(X_{q+1}) & q \geqslant 2 \ . \end{cases}$$

Turning now to BG^+, we of course know that $\pi_1(BG^+) = G/PG$. The homotopy exact sequence for $ABG \to BG \to BG^+$ forces $\pi_q(BG^+) = \pi_{q-1}(ABG)$, $q \geq 3$, and for $q = 2$ reduces to

$$\begin{array}{ccccccccc}
\pi_2(BG) & \longrightarrow & \pi_2(BG^+) & \longrightarrow & \pi_1(ABG) & \longrightarrow & \pi_1(BG) & \twoheadrightarrow & \pi_1(BG^+) \; . \\
\| & & & & \| & & \| & & \| \\
0 & & & & \pi_1(X_3) & & G & & G/PG
\end{array}$$

However, we have already seen that $H_2(X_2) = \mathrm{Ker}\,(\pi_1(X_3) \to G)$.

We now consider in more detail the group extension

$$H_2(PG) \rightarrowtail \pi_1(X_3) \twoheadrightarrow PG$$

arising from the fibration $X_3 = B\pi_1(X_3) \to X_2 = BPG \to K(H_2(PG), 2)$.

(8.2) LEMMA: *An extension* $H' \rightarrowtail H \twoheadrightarrow H''$ *is central if and only if the fibration* $BH' \to BH \to BH''$ *is principal (i.e. induced from* $PK(H', 2) \to K(H', 2)$ *by some map* $BH'' \to K(H', 2)$*).*

Proof. Since $H = \pi_1(BH)$ acts on $\pi_*(BH') = \pi_1(BH') = H'$ by inner automorphisms, the extension is central if and only if the projection $BH \to BH''$ is a simple map. If this projection is induced from the simple map $PK(H', 2) \to K(H', 2)$ then it too is simple. Conversely, if $BH \to BH''$ is known to be simple, then it has a Moore-Postnikov factorization

$$\begin{array}{ccccc}
 & & BH' & & \\
 & & \downarrow & & \\
 & & E & \longrightarrow & PK(H', 2) \\
 & \overset{f}{\nearrow} & \downarrow & \lrcorner & \downarrow \\
BH & \longrightarrow & BH'' & \longrightarrow & K(H', 2)
\end{array}$$

in which f induces an isomorphism of fundamental groups. As all other homotopy groups of BH and E are trivial, this is enough to guarantee that f is a homotopy equivalence (over BH'').

One consequence of (8.2) should be familiar.

(8.3) COROLLARY: *Central extensions of* H'' *with kernel* H' *are classified by* $[BH'', K(H', 2)] \cong H^2(H''; H')$.

More original (in fact it generalizes [14]) is the following application of (6.4) b). (Note that H' abelian implies $BH' = BH'^{+}$ nilpotent.)

(8.4) COROLLARY: *If* $H' \rightarrowtail H \twoheadrightarrow H''$ *is a central extension, then* BH' *is the fibre of* $BH^{+} \to BH''^{+}$.

Now recall from the beginning of Chapter 4 that $\pi_1(X_3) = \pi_1(AX)$ is *superperfect*, which is to say, $H_1(\pi_1(X_3)) = H_2(\pi_1(X_3)) = 0$. (The converse to this remark is now apparent: given a superperfect group S, the acyclic space construction on BS yields in turn $BS = (BS)_2$ and then $BS = (BS)_3$, so that $\pi_1(ABS) = \pi_1((BS)_3) = S$ after all.) So any perfect group P admits a central extension in which it is the image of a superperfect group $SP = \pi_1((BP)_3)$. As before, it follows that $BSP = (BP)_3$. Next, let $H' \rightarrowtail H \twoheadrightarrow P$ be any other central extension of P. After (8.2) we have

$$\begin{array}{ccccc} & \nearrow & BH & \longrightarrow & PK(H', 2) \\ & & \downarrow & & \downarrow \\ BSP & \longrightarrow & BP & \longrightarrow & K(H', 2) \end{array}$$

Just as for (7.3) we obtain an exact sequence

$$[BSP, K(H', 1)] \rightarrowtail [BSP, BH] \longrightarrow [BSP, BP] \longrightarrow [BSP, K(H', 2)]$$

whose extreme terms are $H^i(SP; H')$, $i = 1, 2$, both of which vanish by the universal coefficient theorem [**38** (5.5.3)]. Then there is a unique homotopy class of liftings $BSP \to BH$ and thus (since $\pi_1(Bf) = f : SP = \pi_1(BSP) \to H$) a unique homomorphism $f : SP \to H$ over P. Moreover, for this to occur it is necessary that P be perfect, because for any group J $[BJ, BH] \to [BJ, BP]$ being a monomorphism when $H' = \mathbb{Z}$ implies that J (and hence its image P) is perfect.

(8.5) PROPOSITION: *The category of epimorphisms to a fixed group* P *which have central kernel* (*whose morphisms are homomorphisms over* P) *has an initial object, the* universal central extension *of* P, *if and only if* P *is perfect, in which case the* u.c.e. *of* P *is*

$$H_2(P) \rightarrowtail SP = \pi_1((BP)_3) \twoheadrightarrow P \ .$$

A more group-theoretic discussion of (8.5) may be found in [**31**], where due historical acknowledgement is given. Of course, in its role as the *Schur multiplicator/multiplier* of P, the group $H_2(P)$ has a history of its own (born 1904 – see [**24**]). For a ring A, the *u.c.e.* of the perfect group EA is determined in the next chapter.

To return to the original group G, there $P = PG$ and $BSP = (BP)_3 = (BG)_3$.

(8.6) COROLLARY:

$$\pi_1(BG^+) = G/PG,$$
$$\pi_2(BG^+) = H_2(PG),$$
$$\pi_3(BG^+) = H_3(SPG),$$

where

$$H_2(PG) \rightarrowtail SPG \twoheadrightarrow PG$$

is the u.c.e. of PG.

9 Quillen's higher K-groups

The intention is to define the higher K-groups of a ring A, K_iA ($i \geqslant 1$), as the composition of covariant functors

$$K_i : A \longmapsto GLA \longmapsto BGLA \longmapsto BGLA^+ \longmapsto \pi_i(BGLA^+).$$

As a matter of procedure though, we should first use (8.6) to check that this agrees with the two established cases, K_1 and K_2. The former is quite obvious, since by (1.11) we have that

$$\pi_1(BGLA^+) = GLA/P(GLA) = GLA/EA = K_1A.$$

For the latter, because by (8.6) $\pi_2(BGLA^+)$ is the kernel of the *u.c.e.* of $P(GLA) = EA$, it suffices to show . . .

(9.1) THEOREM: *The central extension* (3.7)

$$K_2A \rightarrowtail StA \twoheadrightarrow EA$$

is the u.c.e. of EA.

Proof. We have to show that $StA = SP(GLA)$, in other words, that StA is superperfect. Certainly $H_1(StA) = 0$ (1.3), leaving only the vanishing of $H_2(StA)$. Now StA is defined as the direct limit of the groups St_nA, and homology preserves direct limits. So we show that, for $n \geqslant 5$, $\iota : St_{n-1}A \to St_nA$ has $H_2(\iota)$ trivial. To do this we shall check that any central extension $K \rightarrowtail G \overset{\psi}{\twoheadrightarrow} St_nA$ splits over $St_{n-1}A$, whence, from (8.3), $H^2(\iota) = 0 : H^2(St_nA; K) \to H^2(St_{n-1}A;K)$. Then by the universal coefficient theorem [38 (5.5.3)] when $K = H_2(St_nA)$,

$$H_2(\iota)^* = 0 : Hom(H_2(St_nA),\ H_2(St_nA)) \to Hom(H_2(St_{n-1}A), H_2(St_nA)),$$

giving the result because $H_2(\iota) = H_2(\iota)^*(id)$.

Note that, since K is central in G, we have, for any $g_1, g_2 \in G$, $[Kg_1, g_2] = [g_1, g_2]$, where we recall that $[g_1, g_2] = g_1 g_2 g_1^{-1} g_2^{-1} = [g_2, g_1]^{-1}$. Three further facts are worth featuring.

In any group G, *for* $u, v, w \in G$,

$$
\begin{array}{lll}
 & & [[u, v], w] \\
(a) & = & [u, v][w, v][v, wu], \\
(b) & = & [[u, v], [w, v]][[w, v], u] \qquad \textit{if } [u, w] \in Z(G), \\
(c) & = & 1 \qquad \textit{if also } [v, w] \in Z(G).
\end{array}
$$

These follow (a) by outright multiplication, and (b) by substitution for $[v, wu] = [v, uw]$ in (a).

Therefore denote by g^a_{ij} the coset $\psi^{-1}(x^a_{ij})$ of K in G ($i, j \leqslant n$; $a \in A$). Define (uniquely) $\sigma : St_{n-1}A \to G$ by means of the function

$$x^a_{ij} \longmapsto [g^1_{in}, g^a_{nj}]$$

on generators (so $i, j < n$). Clearly $\psi \circ \sigma = id$ as required. What has to be checked is that the three sets of relations in $St_{n-1}A$ are respected by σ, in order that it be a homomorphism. First, whenever $i, j, k, \ell \leqslant n$, choose $h \notin \{i, j, k, \ell\}$, $h \leqslant n$ (possible, since $n \geqslant 5$). Then because $g^a_{ij} = K[g^1_{ih}, g^a_{hj}]$ we deduce from (c) above that

$$\text{(d)} \qquad [g^a_{ij}, g^b_{k\ell}] = 1 \quad \textit{if} \quad i \neq \ell,\ j \neq k .$$

As

$$\sigma(1) = \sigma(x^0_{12}) = [g^1_{1n}, g^0_{n2}] = [g^1_{1n}, K] = 1,$$

this implies in particular that, for $i \neq \ell$, $j \neq k$,

$$[\sigma(x^a_{ij}),\ \sigma(x^b_{k\ell})] = 1 = \sigma([x^a_{ij}, x^b_{k\ell}]) .$$

The next argument begins with (a) in the form

$$[v, w][v, u] = [v, wu][w, [v, u]^{-1}] ,$$

and concludes with (d) . . .

$$\sigma(x_{ij}^a)\,\sigma(x_{ij}^b) = [g_{in}^1, g_{nj}^a g_{nj}^b][g_{nj}^a, [g_{in}^1, g_{nj}^b]^{-1}]$$

$$= [g_{in}^1, g_{nj}^{a+b}][g_{nj}^a, g_{ij}^{-b}]$$

(e) $$= \sigma(x_{ij}^{a+b}).$$

Finally, for $i \neq k$, as an application of (b), (d) and (e) in turn, . . .

$$[\sigma(x_{ij}^a), \sigma(x_{jk}^b)] = [[g_{in}^1, g_{nj}^a], g_{jk}^b]$$

$$= [[g_{in}^1, g_{nj}^a], [g_{jk}^b, g_{nj}^a]][[g_{jk}^b, g_{nj}^a], g_{in}^1]$$

$$= [g_{ij}^a, g_{nk}^{-ab}][g_{in}^1, g_{nk}^{-ab}]^{-1}$$

$$= (\sigma(x_{ik}^{-ab}))^{-1}$$

$$= \sigma(x_{ik}^{ab}) .$$

Hence σ preserves the relations in $St_{n-1}A$ and is a homomorphism after all, completing the proof. We remark that with a little extra tedium we could have extended σ over St_nA (proving that St_nA too is superperfect for $n \geqslant 5$); however this would have bogged us down in a check that σ was well-defined over the enlarged domain.

This gives us the "all-clear" to make the longed-for definition:

$$K_iA = \pi_i(BGLA^+), \qquad i \geqslant 1,$$

Not only does (9.1) confirm that $\pi_2(BLA^+)$ is isomorphic to Milnor's K_2A of Chapter 3, but because it also identifies $(BGLA)_3$ as BStA we may apply (8.6) and (7.8).

(9.2) THEOREM: $K_1A = H_1(GLA)$,

$K_2A = H_2(EA)$,

$K_3A = H_3(StA)$,

and

$K_2A \twoheadrightarrow H_2(GLA)$,

$K_3A \twoheadrightarrow H_3(EA)$,

$K_4A \twoheadrightarrow H_4(StA)$.

The last two epimorphisms of (9.2) follow from (7.8), while the surjection $\pi_2(BGLA^+) \twoheadrightarrow H_2(BGLA^+)$ is a consequence of the isomorphism in dimension 1, by the generalized Hurewicz argument again (p. 59).

Now that we have been more forthcoming about the space to which we are applying the plus-construction, we can also be more specific about the construction itself. With extra care we can minimize the number of 2- and 3-cells added to BGLA in (5.1) to form $BGLA^+$. Further, specification of actual attaching maps for these cells gives a determination of the space $BGLA^+$ itself, rather than, as above, of its homotopy type under BGLA. (This of course assumes a definite choice of model for the classifying space, such as Milnor's quotient of the infinite join.) Undoubtedly, this would be a tiresome procedure if it had to be performed separately for each ring A. Fortunately however, the category $\mathbb{R}$ing has an initial object; so once the cells have been adjoined to $BGL\mathbb{Z}$ their addition to BGLA follows from the commuting square

$$\begin{array}{ccc} BGL\mathbb{Z} & \longrightarrow & BGL\mathbb{Z}^+ \\ \downarrow & & \downarrow \\ BGLA & \longrightarrow & BGLA^+ \end{array} \tag{9.3}$$

when it is co-Cartesian, which, by the comment after (5.11), occurs precisely when EA is the normal closure of Im $[E\mathbb{Z} \to EA]$. Thus, as EA is the homomorphic image of StA, the achievement of these goals hinges on the following technical lemma.

(9.4) LEMMA: *In* $St\mathbb{Z}$, *let* $\omega = u_{12}u_{13}^{-1}$ *where* $u_{ij} = x_{ij}^1 x_{ji}^{-1} x_{ij}^1$. *Then* St_nA *is the normal closure of* ω, *provided* $4 \leqslant n \leqslant \infty$.

Proof. The proof is an exercise in StA-manipulation. First show that, for $j \neq 1, k$,

$$u_{1j}x_{1k}^a u_{1j}^{-1} = x_{jk}^{-a} .$$

Thus, for $k \neq 1, 2, 3$,

$$[\omega, x_{3k}^a] = u_{12}x_{2k}^a u_{12}^{-1} x_{3k}^{-a} = x_{2k}^a x_{3k}^{-a} .$$

So (a) of the proof of (9.1) (with $v = x_{h2}^1$, $w = x_{2k}^a$, $u = x_{3k}^{-a}$) implies that

$$[x^1_{h2}, [\omega, x^a_{3k}]] = x^a_{hk} ,$$

provided $h \neq 2, k$. If $h = 2$, then substitution of $v = x^{-1}_{23}$ instead yields that

$$[x^{-1}_{23}, [\omega, x^a_{3k}]] = x^a_{2k}.$$

This means that the normal closure of ω contains x^a_{hk} whenever $k \neq 1, 2, 3, h$. A similar argument shows that it also contains x^a_{kh}; yet for $n \geqslant 4$ every standard generator of St_nA either takes (at least) one of these two forms, or is a commutator of two such.

(9.5) COROLLARY: *The space* $BGLA^+$ *may be chosen so that the square* (9.3) *is co-Cartesian.*

(9.6) COROLLARY: $BGLA^+$ *may be formed from* $BGLA$ *by the adjunction of a single 2-cell and 3-cell to* BGL_3A, *and, for* $n \geqslant 4$, $BGLA^+ = BGLA \cup_{BGL_nA} BGL_nA^+$.

The element u_{12} above was used in the proof of (1.9). Note too that $t = u_{12}$ is conjugate by x^1_{12} to $s = x^{-1}_{21}x^2_{12}$, where s^4 generates $K_2\mathbb{Z}$ (p.26). Moreover, in $E\mathbb{Z}$,

$$\varphi(s^2) = \varphi(t^2) = \begin{pmatrix} -1 & 0 \\ 0 & -1 \end{pmatrix},$$

so that $s^2t^{-2} \in \mathrm{Ker}\,[\varphi : St\mathbb{Z} \to E\mathbb{Z}] = K_2\mathbb{Z} = Z(St\mathbb{Z})$. By putting $u = w = s^2$ and $v = x^1_{12}$ in (a), (c) of the proof of (9.1), we obtain the triviality of $[s^2, x^1_{12}]^2 = (s^2t^{-2})^2$. Hence $s^4t^{-4} = (s^2t^{-2})^2[t^2s^{-2}, t^2]$ is also trivial, making t^4 an alternative description of the generator of $K_2\mathbb{Z}$.

SUMMARY OF RESULTS

Here is a very rapid review of the highlights of what is known about the size of the groups K_iA ($i \geqslant 0$). In general I do not even hint at the methods of proof; suffice it to say they vary widely, both from each other and from those of this book.

In the course of the next two chapters we shall see that $BGLA^+$ has the structure of the space of loops on another space, and therefore admits a multiplication map

$$BGLA^+ \times BGLA^+ \to BGLA^+$$

corresponding to composition of loops. According to Milnor and Moore [32], the rational homotopy type ($\pi_*(X) \otimes \mathbb{Q}$) of such a space, X say, is determined by the rational homology, as follows. An element $x \in H_*(X; \mathbb{Q})$ is said to be *primitive* if

$$\Delta_*(x) = x \otimes 1 + 1 \otimes x,$$

where from (3.11) we recall that $\Delta : X \to X \times X$ is the diagonal map, and $x \mapsto x \otimes 1$ is induced by $in_L : x \mapsto (x, pt)$, etc.

(9.7) $K_iA \otimes \mathbb{Q}$ *is isomorphic to the subgroup of primitive elements of* $H_i(GLA; Q)$.

Here, the homology of $BGLA^+$ has been identified with that of BGLA and so the group GLA. Now let A be the ring of *algebraic integers* (= roots of monic polynomials with integral coefficients) in a *number field* (= finite $\mathbb{Q}$-extension). In this situation, Borel [5] has successfully performed the homological computation. (The statement uses the Kronecker delta again.)

(9.8) *If* A *is the ring of algebraic integers in a number field* F *with* r_1 *real and* r_2 *complex places* (*that is,* $F \otimes_{\mathbb{Q}} \mathbb{R} = \mathbb{R}^{r_1} \times \mathbb{C}^{r_2}$), *then* $K_{2j}A$ *has rank* $\delta_{j,0}$, *and* $K_{2j+1}A$ *has rank* r_2 *for* j *odd and* $r_1+r_2-\delta_{j,0}$ *for* j *even.*

A further constraint on the size of K_iA here was provided by Quillen [41, pp. 179–198].

(9.9) *For* A *as in* (9.8), K_iA *is a finitely generated group.*

So all eyes turn to the order of the finite subgroup of torsion elements in K_iA. In this connection, Lichtenbaum [42, pp. 489–501] has conjectured the ratio of the order of $K_{2j}A$ (j odd) to that of $K_{2j+1}A$ to be (up to 2-torsion) the value of the zeta-function on F at $-j$, when $r_2 = 0$. This conjecture was in part inspired by Quillen's computation in [33] of the K-theory of the finite field $\mathbb{F}_q$.

(9.10) *For* $j \geqslant 1$, $K_{2j}\mathbb{F}_q = 0$, *and* $K_{2j-1}\mathbb{F}_q$ *is cyclic of order* q^j-1.

Impressive support for the conjecture in general has been provided by Harris and Segal [16]. (See also Browder [8].) However, the specific case to arouse most excitement is of course that where $A = \mathbb{Z}$, $F = \mathbb{Q}$. Now the zeta-function at $-(2n-1)$ assumes the value $-B_n/2n$, the *Bernoulli number* B_n being (2n)! times the coefficient of t^{2n} in the power series expansion of $t/(e^t-1)$.

More ambitious versions of the conjecture relate the orders of $K_{4n-2}\mathbb{Z}$, $K_{4n-1}\mathbb{Z}$ to the numerator and denominator respectively of this fraction. Soulé [37] has obtained partial confirmation here, and his results have recently been extended by Dwyer and Friedlander (to appear).

The denominator of $B_n/4n$ is familiar to homotopy-theorists as the order of the (cyclic) image of the J-homomorphism in homotopy groups of spheres, which image Quillen [43, pp. 182–188] shows maps injectively to the homotopy groups of $BGL\mathbb{Z}^+$. Hence $K_{4n-1}\mathbb{Z}$ contains an element of order equal to the denominator of $B_n/4n$ ($= \frac{1}{24}$ for $n = 1$, $\frac{1}{240}$ for $n = 2$, etc.). Moreover, this generates a cyclic summand when n is even. For n odd, it need not generate a summand, as Lee and Sczczarba revealed in their paper. *The group* $K_3(\mathbb{Z})$ *is cyclic of order forty-eight* [27]. Further torsion results appear in [7], [28], [36], [37]; however, the comparison with knowledge of stable homotopy groups of spheres points up how rudimentary is our understanding of $K_*\mathbb{Z}$.

10 Delooping of the plus construction

The purpose of this chapter is to instal the machinery which shows that $BGLA^+$ is an infinite loop space. Thus the space T_1 in this chapter becomes BGLA in the next.

Note from (7.2) that any map $X \to U$ gives rise, for $1 \leqslant n \leqslant \infty$, to a map $p_{n+1} : AX \to U_{n+1}$ and thus a fibre sequence $F_{p_{n+1}} \to AX \to U_{n+1}$, although in general the homotopy fibre $F_{p_{n+1}}$ is difficult to identify. If however one of the conditions of (6.4) is satisfied, then we deduce from the fibre sequence $F^+_{p_{n+1}} \to * \to U^+_{n+1}$ that $F^+_{p_{n+1}}$ is (n−1)-connected, so that $\widetilde{H}_i(F_{p_{n+1}}) = \widetilde{H}_i(F^+_{p_{n+1}}) = 0$ for $i < n$, making $F_{p_{n+1}}$ the n-th stage of some acyclic tower (*cf.* (7.1)). This is the case we now examine.

(10.1) LEMMA: *Suppose that* T, O, U *are spaces,* O *acyclic, such that, for some* $n \geqslant 1$, $T_n \to O \xrightarrow{p_{n+1}} U_{n+1}$ *is a fibre sequence. Then the three conditions* –

(i) p_{n+1} *is quasi-nilpotent (and* T^+_n *is nilpotent if* $n = 1$),

(ii) $q_n : T_n \to \Omega U^+_{n+1}$ *(defined below) is acyclic,*

(iii) $p_\infty : O \to AU$ *is acyclic* –

are equivalent and imply that $T_{n+1} \to O \xrightarrow{p_{n+2}} U_{n+2}$ *is also a fibre sequence.*

Proof. Effectively, we show that there is a commuting diagram

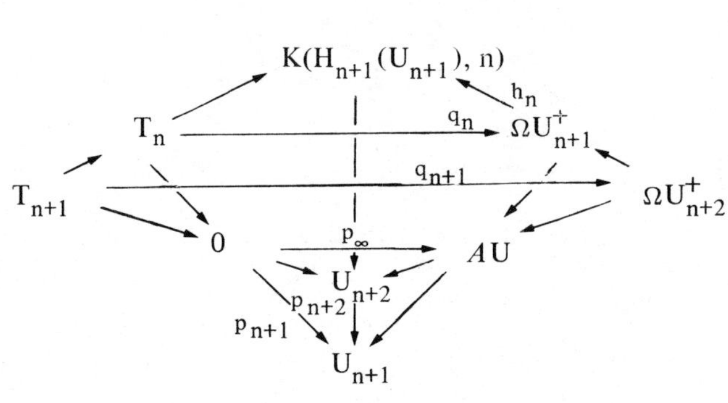

in which all collineations are fibre sequences, all horizontal maps acyclic. q_n is the map of homotopy fibres induced from $O \to AU$ over U_{n+1}. (Make what you will of the fact that this is Desargues' configuration with the common fibre (4.2) of q_n, q_{n+1}, p_∞ – identified by (10.2) as AT – omitted/at infinity.)

For the proof in earnest, begin by observing that the equivalence of (ii), (iii) follows from (4.2). To see that each is a consequence of (i), observe that p_∞ is a homology equivalence because $\widetilde{H}_*(O) = \widetilde{H}_*(AU) = 0$. So, for $n \geqslant 2$, (iii) comes via (4.13) b), ΩU^+_{n+1} being (n–1)-connected (7.8). For n = 1 we apply (4.4) ($AU \to U_2$ being quasi-nilpotent after (7.5)) to deduce that q_1, and hence $q_1^+ : T_1^+ \to \Omega U_2^+$, is a homology equivalence. Since both T_1^+ and ΩU_2^+ are nilpotent (the latter, *e.g.*, by (7.5), (4.8)), this forces q_1^+ to be a homotopy equivalence (4.18) and q_1 itself to be acyclic.

Conversely, because the action of $\pi_1(U_{n+1})$ on $H_*(T_n)$ corresponds through $H_*(q_n)$ to that on $H_*(\Omega U^+_{n+1})$, (ii) implies (i). (When n = 1, (ii) forces (after (5.1)) T_1^+ to be homotopy equivalent to the nilpotent space ΩU_2^+.)

We use condition (ii) for the remaining argument. From the fibre sequence

$$\Omega U^+_{n+2} \longrightarrow \Omega U^+_{n+1} \xrightarrow{h_n} K(H_{n+1}(U_{n+1}), n) \longrightarrow U^+_{n+2}$$

(an application of (6.4)), $\pi_i(h_n)$ is an isomorphism whenever $i \leqslant n$, so that, after Hurewicz, $H_i(h_n)$ is too. For n = 1, $\pi_1(\Omega U_2^+) = \pi_2(U_2^+) = H_2(U_2)$ (7.8) is abelian, so that by (5.1) acyclic q_1 is just $q_T : T \to T^+$ (up to homotopy under T), and so $T \xrightarrow{h_1 \circ q_1} K(H_2(U_2), 1)$ is just $T \to K(\pi_1(T)/P\pi_1(T), 1)$, with fibre T_2. For $n \geqslant 2$, $T_n \xrightarrow{h_n \circ q_n} K(H_{n+1}(U_{n+1}), n)$ is an isomorphism on n-th homology groups and thus (up to homotopy under T_n) is the map $T_n \to K(H_n(T_n), n)$, with fibre T_{n+1} (*cf.* (7.1)). In either event,

$$T_{n+1} \simeq \text{fibre } [T_n \xrightarrow{Fp_{n+2}} K(H_{n+1}(U_{n+1}), n)] = F_{Fp_{n+2}} \simeq F_{p_{n+2}} .$$

The fact that condition (iii) of (10.1) is independent of n makes iteration possible. By applying either (6.4) b) to (i) or (5.1) to (ii) we have

(10.2) THEOREM: *If* T *is the homotopy fibre of quasi-nilpotent* $p_2 : O \to U_2$ *with* O *acyclic,* T^+ *nilpotent, then for all* $n \geqslant 1$, $q_n^+ : T_n^+ \to \Omega U^+_{n+1}$ *is a homotopy equivalence.*

We leave it to the reader as an exercise (using (6.4) a), (4.10)) to check that. . .

(10.3) T^+ *is nilpotent if and only if (for* $\pi = \pi_1(T)$) $\pi/P\pi$ *is nilpotent and the fibration* $T_2 \to T \to K(\pi/P\pi, 1)$ *is quasi-nilpotent.*

The most exciting case of (10.2) occurs when T, U are closely related to one another.

(10.4) COROLLARY: *Suppose* C *is a category and* $\mathbf{S} : C \to C$, $\mathbf{T} : C \to \mathbb{T}\mathrm{op}$ *(our topological category) are such that, for any object* A *in* C, *there is an acyclic space* O_A *and quasi-nilpotent fibration* $\mathbf{T}(A) \to O_A \to (\mathbf{T}(\mathbf{S}A))_2$ *with* $\mathbf{T}(A)^+$ *nilpotent. Then there is (functorially) an* Ω*-spectrum*

$$\ldots, \Omega^2\mathbf{T}(A)^+, \Omega\mathbf{T}(A)^+, \ \mathbf{T}(A)^+, \ \mathbf{T}(\mathbf{S}A)_2^+, \ \mathbf{T}(\mathbf{S}^2A)_3^+, \ldots$$

In particular, $\mathbf{T}(A)^+$ *is an infinite loop space.*

The content of this result is that, given a morphism $f : A \to A'$ in C, then for any $n \geqslant 1$ there is a diagram

$$\begin{array}{ccc} (\mathbf{TS}^{n-1}A)_n^+ & \xrightarrow{q_n^+} & \Omega(\mathbf{TS}^nA)_{n+1}^+ \\ \downarrow & & \downarrow \\ (\mathbf{TS}^{n-1}A')_n^+ & \xrightarrow{q_n^+} & \Omega(\mathbf{TS}^nA')_{n+1}^+ \end{array}$$

commuting (after (5.2), (7.1)) up to homotopy under $(\mathbf{TS}^{n-1}A)_n$, in which each q_n^+ is a homotopy equivalence. So the n-th term, $(\mathbf{TS}^nA)_{n+1}^+$, is the n-th delooping of the space $\mathbf{T}A^+$. We would thus be justified in writing $\mathbf{T}A^+$ as $\Omega^\infty(A\mathbf{TS}^\infty A^+)$, just as homotopy-theorists write $\Omega^\infty S^\infty$ although the infinite-dimensional sphere S^∞, like $A\mathbf{TS}^\infty A^+$, is contractible. *Infinite loop spaces* are investigated in J.F. Adams' Annals of Math. Study (no. 90) of that name, whose first chapter relates them to Ω-spectra and cohomologies. For the present we merely remark that an Ω-spectrum $\mathbf{E} = \{E_{(r)}\}_{r \in \mathbb{Z}}$ (*i.e.* $E_{(r)} \simeq \Omega E_{(r+1)}$, $r \geqslant 0$) gives rise to a *generalized cohomology theory* $h^*(\ ;\mathbf{E})$ by means of the assignments (X compact)

$$X \mapsto h^n(X; \mathbf{E}) \ = \ [X \sqcup pt, E_{(n)}].$$

It is said to be p-*connected* if $h^{-n}(pt; \mathbf{E}) = 0$ whenever $n \leqslant p$, or equally, if $E_{(r)}$ is (r+p)-connected whenever $r+p \geqslant 0$. Thus our spectrum is 0-connected (7.8). It can be amended to a (−m)-connected spectrum ($0 \leqslant m \leqslant \infty$) with unchanged positive homotopy groups

$$h^{-n}(pt; \mathbf{E}) = [S^0, E_{(-n)}] = \pi_n(E_{(0)}), \qquad n > 0,$$

where

$$(10.5) \qquad E_{(r)} = \begin{cases} \mathbf{T}S^r A^+_{r-m+1} & r \geqslant m, \\ \mathbf{T}S^r A^+ \times \pi_1(\mathbf{T}S^{r+1} A^+) & 0 \leqslant r < m, \\ \Omega^{-r}(\mathbf{T}A^+ \times \pi_1(\mathbf{T}SA^+)) & r < 0. \end{cases}$$

However, it has to be said that a clear advantage of at least (−1)-connected (= *connective*) spectra, such as in (10.4), is the attractive uniqueness property they possess **[13]**, **[30]**. In the present circumstances it is handy to have a comparison between the 0-connected and (−1)-connected spectra (written $\overline{\mathbf{T}}\mathbf{A}$ and $\mathbf{TA}$ respectively). To obtain this, recall the fibre sequence

$$\mathbf{T}S^r A^+_{r+1} \to \mathbf{T}S^r A^+_r \to K(H_r(\mathbf{T}S^r A_r), r)$$

used in (7.8). In the usual way the sequence extends to the left by successive "looping" of each space in turn. When (10.4) applies, it also extends to the right by "delooping" (since too $K(H, r) = \Omega K(H, r+1)$). Now

$$H_r(\mathbf{T}S^r A^+_r) \cong \pi_r(\mathbf{T}S^r A^+_r) = \pi_1(\Omega^{r-1}(\mathbf{T}S^{r-1} SA^+_r)) = \pi_1(\mathbf{T}SA^+).$$

So when we consider homotopy classes of maps from a space X (with disjoint basepoint added) into the spaces of the extended fibre sequence

$$\mathbf{T}A^+ \to \mathbf{T}A^+ \times \pi \to \pi \to \mathbf{T}SA^+_2 \to \mathbf{T}SA^+ \to K(\pi, 1) \to \mathbf{T}S^2A^+_3 \to \mathbf{T}S^2A^+_2 \to \ldots$$

where $\pi = \pi_1(\mathbf{T}SA^+)$, we obtain the long exact sequence

$$(10.6) \qquad h^0(X; \overline{\mathbf{T}}\mathbf{A}) \to h^0(X; \mathbf{TA}) \to H^0(X; \pi_1(\mathbf{T}SA^+)) \to h^1(X; \overline{\mathbf{T}}\mathbf{A}) \to \ldots$$

Clearly, the first three terms of (10.6) form a split short exact sequence, comparable to (2.5).

One advantage of cohomology theories over homology theories in general is that they often carry a (graded) *multiplicative* structure. This arises from a collection of product maps, aggregating to a *product of spectra*

$$E_{(m)} \wedge E_{(n)} \to E_{(m+n)} \qquad (m, n \geqslant 0) .$$

To achieve this, we apply (7.4) and (5.8) to our 0-connected spectrum $\overline{\mathbf{T}}\mathbf{A}$.

(10.7) COROLLARY: *In the context of* Corollary 10.4, *functorial maps*

$$\mathbf{TS}^m\mathrm{A} \wedge \mathbf{TS}^n\mathrm{A}' \to \mathbf{TS}^{m+n}\mathrm{A}''$$

induce a functorial product of spectra

$$(\mathbf{TS}^m\mathrm{A})^+_{m+1} \wedge (\mathbf{TS}^n\mathrm{A}')^+_{n+1} \to (\mathbf{TS}^{m+n}\mathrm{A}'')^+_{m+n+1} \ .$$

It is not hard to see what further structure one should demand in order to ensure associativity and commutativity, where appropriate. Our interest of course lies in the case where the category *C* is $\mathbb{R}$ing, which is where the next chapter takes up the story.

11 The algebraic K-theory spectrum

In this chapter we apply the "de-looping machine" (10.4) to the key space $BGLA^+$. We thus have to make a suitable choice for functors **S**, **T** and then show that the conditions of (10.4) are satisfied. Choice of **T** : $\mathbb{R}$ing → $\mathbb{T}$op is obvious: we just compose the general linear group functor GL : $\mathbb{R}$ing → $\mathbb{G}$roup and the classifying space functor B : $\mathbb{G}$roup → $\mathbb{T}$op.

S however requires some invention. Various descriptions are available; for our purposes the most convenient is the following, borrowed from Chapter 3. Recall from Chapter 1 that the pseudo-ring mA of finite matrices serves as a copy of A inasmuch as $GLmA \cong GLA$ (1.14). Now mA is a two-sided ideal in the ring CA of locally-finite matrices, entries in A (addition and multiplication as for finite matrices); such matrices have only a finite number of non-zero entries in each row and column. We further insist that the elements of a matrix in CA be drawn from a finite sample of elements of A. This extra condition, by ensuring that $CA \cong C\mathbb{Z} \otimes_{\mathbb{Z}} A$, is invaluable whenever the construction is iterated. (However for present purposes it is quite redundant: it follows from the results below that even when it is omitted the same space $BGLCA^+$ ensues.) In fact mA is the extended ideal A^e under the inclusion $A \hookrightarrow CA$

$$a \mapsto \left(\begin{array}{c|c} a & 0 \dots \\ \hline 0 & \\ \vdots & 0 \end{array}\right).$$

Therefore define SA to be the quotient ring CA/mA. Since the ring homomorphism $Cf : CA_1 \to CA_2$ induced by a homomorphism $f : A_1 \to A_2$ sends mA_1 to mA_2, the construction S is indeed functorial and serves as this chapter's example for **S** : $\mathbb{R}$ing → $\mathbb{R}$ing. Again, for later convenience we note that there is a natural isomorphism between the functors $A \mapsto SA$ and $A \mapsto S\mathbb{Z} \otimes_{\mathbb{Z}} A$.

With these choices for **S**, **T**, the theorem dictated by (10.4) is as follows.

(11.1) THEOREM: *There is a functor from* Ring *to the category of* 0*-connected* Ω*-spectra which associates to a ring* A *the* Ω*-spectrum* $\overline{\mathbf{K}}\mathbf{A}$, *where* $\overline{\mathrm{K}}\mathrm{A}_{(r)}$ *is defined as* $(\mathrm{BGLS^rA})^+_{r+1}$ *for* $r \geqslant 0$ *and* $\Omega^{-r}(\mathrm{BGLA}^+)$ *for* $r < 0$.

This result immediately enables us to compare the higher K-groups of a ring with those of its suspension. By putting n = 0 (for example), we obtain that for all $i \geqslant 1$,

$$K_iA = \pi_i(\mathrm{BGLA}^+) \cong \pi_i(\Omega(\mathrm{BGLSA})^+_2) = \pi_{i+1}((\mathrm{BGLSA})^+_2).$$

However, after (7.8) this last group is just $\pi_{i+1}(\mathrm{BGLSA}^+) = K_{i+1}(\mathrm{SA})$.

(11.2) COROLLARY: *For* $i \geqslant 1$, $K_iA \cong K_{i+1}SA$.

This is the desired extension of (3.3). Again, it serves to validate the present definition of higher K-groups. We may call $\overline{\mathbf{K}}\mathbf{A}$ the 0*-connected K-theory spectrum*. The corresponding (−1)-connected spectrum, **KA**, is (after (10.5), (3.3) b))

$$\mathrm{KA}_{(r)} = \begin{cases} (\mathrm{BGLS^rA})^+_r & r \geqslant 1\ , \\ \Omega^{-r}(K_0A \times \mathrm{BGLA}^+) & r \leqslant 0\ . \end{cases}$$

Now (10.6) links the cohomology theories which derive from these spectra. With the notation $\overline{\mathrm{K}}\mathrm{A}^n(X) = h^n(X; \overline{\mathbf{K}}\mathbf{A}) = [X, (\mathrm{BGLS^nA})^+_{n+1}]$, etc., its output is as follows.

(11.3) COROLLARY: *For a compact CW-complex* X, *the (functorial) sequence*

$$\overline{\mathrm{K}}\mathrm{A}^0(X) \to \mathrm{KA}^0(X) \to H^0(X; K_0A) \to \overline{\mathrm{K}}\mathrm{A}^1(X) \to \mathrm{KA}^1(X) \to H^1(X; K_0A) \to \dots$$

is exact, with the first three terms constituting a split short exact sequence.

To return to (11.1) itself, the spaces $(\mathrm{BGLSA})_2$ and $(\mathrm{BGLS^2A})_3$ appearing there have already been identified in a more algebraic way as BESA and $\mathrm{BStS^2A}$ respectively (9.2), a point we pursue below (11.12). The immediate task is to check the requisite hypotheses for (10.4). Specifically we prove in turn . . .

(11.4) *There is a fibration*

$$\mathrm{BGLA} \longrightarrow \mathrm{BGLCA} \longrightarrow \mathrm{BESA}.$$

(11.5) *The fibration* (11.4) *is quasi-nilpotent.*

(11.6) $BGLA^+$ *is nilpotent.*

(11.7) BGLCA *is acyclic.*

The proof of (11.4) is very quick. We just take the exact sequences of groups

$$1 \to GLmA \to GLCA \to GLSA, \qquad ECA \to ESA \to 1$$

which result from the definition of SA, borrow the facts that GLmA $\cong$ GLA and (from (11.7)) that ECA = GLCA, and apply the classifying space functor.

For (11.5) and (11.6) we need to recall, as we did for (3.11), a key fact from group homology theory: an inner automorphism of a group induces the identity automorphism on its homology. Now homology preserves direct limits, and any group is the direct limit of its finitely generated subgroups. So a group automorphism which acts on each finitely generated subgroup as an inner automorphism of that subgroup again acts trivially on the group's homology. Thus to prove that one group acts trivially on the homology of another it suffices to find, for each finite subset of the latter, an inner automorphism of the latter to which the restricted action corresponds. Perhaps the simplest, feasible guarantee of success is the hypothesis of the following lemma (wherein $C_G(H)$ denotes the centralizer in G of a subgroup H).

(11.8) LEMMA: *Suppose* $\varinjlim N_\alpha = N \trianglelefteq G$ *where, for all* α, $N_\alpha \leqslant G$ *and*

$$G = N \,.\, C_G(N_\alpha).$$

Then G (*and thereby* G/N) *acts trivially on* $H_*(N)$.

Proof. An arbitrary finite subset of N is contained in some N_α. For this α a given element $g \in G$ may be written as

$$g = g_1 g_2 \qquad g_1 \in N, \; g_2 \in C_G(N_\alpha) \;.$$

Then for all $h \in N_\alpha$

$$ghg^{-1} = g_1(g_2hg_2^{-1})g_1^{-1} = g_1hg_1^{-1} \;.$$

It is amusing to note that the particular case where G can be written in the form $G = N_\alpha C_G(N_\alpha)$, which occurs in the proof of (11.5) for example, may also be deduced from (3.11). One attraction of the hypothesis of (11.8) is its iterative potential . . .

(11.9) COROLLARY: *Suppose* $\varinjlim K_\beta = K \trianglelefteq \varinjlim N_\alpha = N \trianglelefteq G$ *where, for all* α, β, *there holds* $N_\alpha \leqslant G$, $K_\beta \leqslant N$ *and*

$$G = N \,.\, C_G(N_\alpha) \qquad \textit{and} \qquad N = K \,.\, C_N(K_\beta) \;.$$

Then G *acts trivially on* $H_*(K)$.

For completeness we should also include the corresponding quasi-nilpotence result. Let $\Gamma_G^i N$ be the lower central series of a normal subgroup N ($\Gamma_G^0 N = N$, $\Gamma_G^{i+1} N = [G, \Gamma_G^i N]$). The definition that G act nilpotently on N reduces to $\Gamma_G^k N = 1$ for some k, or, equivalently, to $N \leqslant \zeta_k(G)$ for the same k ($\zeta_k(G)$ as in (1.6) – see p. 37).

(11.10) *Suppose* $N \trianglelefteq G$ *and for some* k $\Gamma_G^k N = \varinjlim H_\alpha$ *where, for all* α, $H_\alpha \leqslant G$ *and*

$$G = \Gamma_G^k N \,.\, C_G(H_\alpha) \;.$$

Then G *acts nilpotently on* $H_*(N)$.

The proof applies [22 Theorem 2.4] to the G-action on the quasi-nilpotent (after (11.8)) fibration

$$B\Gamma_G^k N \longrightarrow BN \longrightarrow B(N/\Gamma_G^k N) \;.$$

(By definition, G acts nilpotently on $N/\Gamma_G^k N$, and hence on its homology.) It's not necessary to spell out the proof because in the cases of interest to us N is perfect so that always $\Gamma_G^i N = N$ and (11.8) already applies.

Returning then to these relevant cases, we prove (11.5) by showing (11.8) applies to $N = GLmA = \varinjlim GLM_nA$, $G = GLCA$. Of course it suffices to consider only generators of G; thanks to the by-product GLCA = ECA of (11.7) these are elementary matrices of the form e_{rs}^γ where $\gamma \in CA$. Now, for each n there is a decomposition (Ann = annihilator ideal)

$$\gamma = \gamma_1 + \gamma_2 \in M_nA + \mathrm{Ann}(M_nA),$$

or equally

$$e^{\gamma}_{rs} = e^{\gamma_1}_{rs} \cdot e^{\gamma_2}_{rs} \in GLM_nA \cdot C_{GLCA}(GLM_nA) .$$

Likewise, for the action of GLA on $H_*(EA)$, let $g \in GLA$ and n be given. Write $h = \max(n, k)$ where $g \in GL_kA$, and thus

$$g = \begin{pmatrix} \alpha & 0 \\ 0 & I \end{pmatrix}$$

with $\alpha \in M_hA$. Now (from (1.9))

$$\begin{aligned} g &= \begin{pmatrix} \alpha & 0 & 0 \\ 0 & \alpha^{-1} & 0 \\ 0 & 0 & I \end{pmatrix} \begin{pmatrix} I_h & 0 & 0 \\ 0 & \alpha & 0 \\ 0 & 0 & I \end{pmatrix} \\ &\in E_{2h}A \cdot C_{GLA}(GL_hA) \\ &\leqslant EA \cdot C_{GLA}(E_nA) . \end{aligned}$$

The action on homology is thus trivial, making the fibration

$$BEA \longrightarrow BGLA \longrightarrow K(GLA/EA, 1)$$

quasi-nilpotent. Since $GLA/EA = GLA/PGLA = GLA_{ab}$ (1.11) is certainly a nilpotent group, (10.3) reveals $BGLA^+$ to be a nilpotent space, verifying (11.6).

This leaves the key result, (11.7), which we prove by showing GLCA conforms to the following model for acyclic (= homologically trivial) groups.

(11.11) LEMMA: *Suppose a group* G *is the direct limit of subgroups* G_λ, $\lambda \in \Lambda$, *each equipped with a homomorphism* $\phi_\lambda : G_\lambda \to C_{G_\mu}(G_\lambda)$, $\lambda \leqslant \mu = \mu_\lambda \in \Lambda$, *and an element* $a \in G_\mu$ *such that, for all* $g \in G_\lambda$, $g = [a, \phi_\lambda(g)]$. *Then* G *is acyclic.*

The above lemma is an immediate corollary of (3.11), obtained by letting $\rho : G \to G$ be the trivial endomorphism.

To complete the proof of (11.7), and thereby (11.1), we must show that $GLCA = \varinjlim GL_nCA$ fits the framework of this lemma. Crucial here is the existence of a bijection, say $\gamma : \mathbb{N} \to \mathbb{N} \times \mathbb{N}$.

For this induces an isomorphism $\gamma^* : A^{\mathbb{N}\times\mathbb{N}} \to A^{\mathbb{N}}$ which may be regarded as a re-indexing of bases; thus our original ring CA defined on the countably generated free module $A^{\mathbb{N}}$ is isomorphic to the ring $\hat{C}A$ of matrices which are locally finite with respect to the module $A^{\mathbb{N}\times\mathbb{N}}$ via $CA \to \hat{C}A, \alpha \mapsto \gamma^{*-1} \circ \alpha \circ \gamma^* = \hat{\alpha}$. There is then defined a homomorphism

$$\psi : \hat{C}A \to \hat{C}A$$

$$\hat{\alpha} \mapsto [\alpha \oplus \alpha \oplus \alpha \oplus \ldots : \underset{\mathbb{N}}{\oplus} A^{\mathbb{N}} \to \underset{\mathbb{N}}{\oplus} A^{\mathbb{N}}] .$$

It is in fact $\psi_* : GL_n(\hat{C}A) \to GL_n(\hat{C}A)$ which we use, to define

$$\phi_n : GL_n(\hat{C}A) \to GL_{3n}(\hat{C}A)$$

$$\hat{\alpha} \mapsto \begin{pmatrix} I_n & 0 & 0 \\ 0 & \psi_*\hat{\alpha} & 0 \\ 0 & 0 & I_n \end{pmatrix}$$

Evidently $[GL_n(\hat{C}A), \phi GL_n(\hat{C}A)] = 1$, so that it remains to identify a matrix $\mu_n \in GL_{3n}(\hat{C}A)$ with the property that $\hat{\alpha} = [\mu_n, \phi_n(\hat{\alpha})]$. Since it is actually a permutation matrix which does the trick, the easiest way to describe it is as a permutation of the basis elements, which we write as ${}^h({}_ie^k)_\ell = {}^h_i(e^k)_\ell = {}^h_ie^k_\ell \in \bigoplus_{h=1}^{3} \bigoplus_{i=1}^{n} \bigoplus_{\ell\in\mathbb{N}} A^{\mathbb{N}}$.

Just to check this notation, observe that $\phi_n(\hat{\alpha})$ above behaves as

$${}^h_ie^k_\ell \mapsto \begin{cases} {}^1_ie^k_\ell & h = 1 , \\ {}^2(\alpha({}_ie^k))_\ell & h = 2 , \\ {}^3_ie^k_\ell & h = 3 . \end{cases}$$

Now define μ_n by

$${}^h_ie^k_\ell \quad \begin{cases} {}^3_i(\gamma^*e^k_\ell)_1 & h = 1 , \\ {}^1_i(\gamma^{*-1}e^k) & h = 2,\ \ell = 1 , \\ {}^2_ie^k_{\ell-1} & h = 2,\ \ell > 1 , \\ {}^3_ie^k_{\ell+1} & h = 3 . \end{cases}$$

The picture of μ (for each i) is rather pretty:–

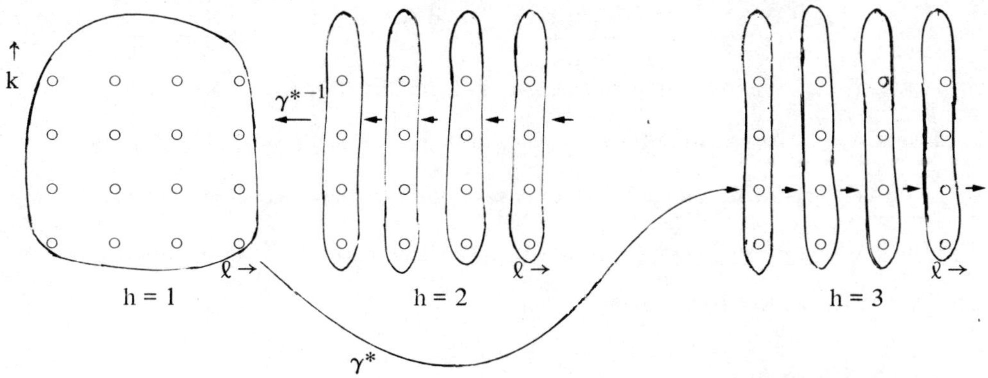

It is (genuinely) easy to verify that these two matrices have commutator $\hat{\alpha}$ after all. So we are done.

There is a bonus, inasmuch as both μ_n is always a permutation matrix and $\phi_n(\hat{\alpha})$ is, so long as $\hat{\alpha}$ was. Thus the lemma also establishes the acyclicity of the group of infinite permutations of $\mathbb{N} \times \mathbb{N}$ which leave all but finitely many columns (copies of $\mathbb{N}$) invariant. In fact the lemma serves to establish the acyclicity of many other groups besides, but that, as they say, is another story . . .

Finally, if an infinite delooping of $BGLA^+$ seems like a concept without much algebraic motivation, I can offer the following interpretation of the second stage. The first, you will recall, came from the short exact sequence

$$GLA \rightarrowtail GLCA = ECA \twoheadrightarrow ESA .$$

Had we applied (10.1) instead of the more ambitious (10.4), we'd have come across a fibration $BEA \rightarrow BGLCA \rightarrow BStA$ (since $BStA = (BGLA)_3$ (9.2)). Historically, I only suspected that something like (10.1) might hold simply because I had first established the purely algebraic result . . .

(11.12) *There is a short exact sequence*

$$EA \rightarrowtail ECA = StCA \twoheadrightarrow StSA .$$

Since a ring epimorphism must induce an epimorphism of Steinberg groups, the proof consists in identifying $K = \mathrm{Ker}\,[StCA \rightarrow StSA]$. As a first step, note that

$$K \leqslant \operatorname{Ker}\,[ECA = StCA \to StSA \to ESA] \;=\; GLA\,.$$

To obtain a lower bound for K we argue in the converse direction: because $EA \leqslant GLA$ the homomorphism $EA \to ECA \to StSA \to ESA$ is trivial, so that $EA \to StSA$ must factor through $\operatorname{Ker}\,[StSA \to ESA] = H_2(ESA)$. Actually, we only need the fact that this kernel is abelian as this implies the triviality of the image of any perfect group, such as EA. Thus the original homomorphism $EA \to StSA$ is also trivial, leaving

$$EA \leqslant K \leqslant GLA\ .$$

The proof is completed by two-fold application of the exact sequence

$$H_2(G) \to H_2(Q) \to N/[G, N] \to G_{ab} \to Q_{ab}$$

arising classically from a short exact sequence $N \hookrightarrow G \twoheadrightarrow Q$. In the first instance, from $K \hookrightarrow ECA \twoheadrightarrow StSA$ we deduce, because ECA is perfect while StSA is superperfect, that $K = [ECA, K]$. In the second, from $GLA \hookrightarrow ECA \twoheadrightarrow ESA$ we have, ECA being superperfect, that

$$\begin{aligned} GLA/[ECA, GLA] &\cong H_2(ESA) \\ &= K_2(SA) && \text{(by (9.2))} \\ &= K_1(A) && \text{(by (3.3) c))} \\ &= GLA/EA\,. \end{aligned}$$

Since $EA = [GLA, GLA] \leqslant [ECA, GLA]$, it must be that

$$EA \;=\; [ECA, GLA]\ .$$

This combines with the previous pieces of information to reveal that

$$EA \leqslant K \;=\; [ECA, K] \leqslant [ECA, GLA] \;=\; EA,$$

and hence $K = EA$ after all.

12 Change of rings

This chapter considers two aspects of a ring homomorphism $f : A \to A_1$. In the first place we attempt a ring-theoretic description of the fibre of the map induced by f on $BGLA^+$, so as to give the associated exact homotopy sequence further algebraic content. Secondly, when f is (very) well-behaved, we describe a technique for having it determine a contravariant transfer map from K_iA_1 to K_iA.

The idea for constructing a ring-theoretic fibre for f (expounded in [**40**]) is inspired, unsurprisingly, by the topological procedure. This involves a canonical fibration π from a homotopically trivial space (the path space, PX_1) to the target space X_1 and taking the pull-back

$$\begin{array}{ccc} F_g & \longrightarrow & X \\ \downarrow & \lrcorner & \downarrow g \\ PX_1 & \xrightarrow[\pi]{} & X_1 \end{array} \quad .$$

In the previous chapter we obtained a convincing model for a K-theoretically trivial ring, namely CA_1. Unfortunately this does not map usefully to A_1, so that we're forced to tensor with $S\mathbb{Z}$ and take instead as our fibre the pull-back

$$\begin{array}{ccc} R_f & \longrightarrow & SA \\ \downarrow & \lrcorner & \downarrow Sf \\ CA_1 & \twoheadrightarrow & SA_1 \end{array} \quad .$$

This means that $GLR_f = GLCA_1 \times_{GLSA_1} GLSA$. However, we have already seen (*e.g.* (11.1), (11.4)) that it is more profitable to deal with ESA rather than GLSA. So define the group $\bar{E}$ to be the fibre-product $ECA_1 \times_{ESA_1} ESA$ $(= GLCA_1 \times_{GLSA_1} ESA)$. Evidently, $ER_f \leqslant \bar{E} \leqslant GLR_f$. So, as an immediate consequence of (1.11), we see that $ER_f = P\bar{E}$, the maximal perfect subgroup of $\bar{E}$. Hence, from (7.8), we obtain the justification for studying the space $B\bar{E}^+$ instead of $BGLR_f^+$.

(12.1) *For* $q \leqslant 2$, $K_q R_f = \pi_q(B\bar{E}^+)$.

The part played by the space $B\bar{E}^+$ comes from the induced group extension

$$\begin{array}{ccccc} GLA_1 & \rightarrowtail & \bar{E} & \twoheadrightarrow & ESA \\ \downarrow{\scriptstyle id} & & \downarrow & \lrcorner & \downarrow{\scriptstyle ESf} \\ GLA_1 & \rightarrowtail & ECA_1 & \twoheadrightarrow & ESA_1 \end{array} .$$

Note that the action of ESA on $H_*(GLA_1)$, being induced from the action of ESA_1, is trivial, after the proof of (11.5). Then (11.6) makes (6.4) b) applicable, leading to the fibrations

$$\begin{array}{ccccc} BGLA_1^+ & \longrightarrow & B\bar{E}^+ & \longrightarrow & BESA^+ \\ \downarrow{\scriptstyle id} & & \downarrow & & \downarrow{\scriptstyle BESf^+} \\ BGLA_1^+ & \longrightarrow & BECA_1^+ & \longrightarrow & BESA_1^+ \end{array} .$$

Now $BECA_1^+$ is contractible because $BECA_1$ is acyclic (11.7). So $BECA_1^+ \times_{BESA_1^+} BESA^+$ serves as a model for the homotopy fibre, F, say, of $BESf^+ : BESA^+ \to BESA_1^+$. The resulting map of fibrations over $BESA^+$,

$$\begin{array}{ccccc} BGLA_1^+ & \longrightarrow & B\bar{E}^+ & \longrightarrow & BESA^+ \\ \downarrow{\scriptstyle id} & & \downarrow & & \downarrow{\scriptstyle id} \\ BGLA_1^+ & \longrightarrow & F & \longrightarrow & BESA^+ \end{array} ,$$

ensures $B\bar{E}^+$ has the right homotopy type for the fibre of $BESf^+$. Hence, in passing to the homotopy exact sequence for $BESf^+$, we obtain

$$\ldots \to \pi_{q+1}(BESA_1^+) \to \pi_q(B\bar{E}^+) \to \pi_q(BESA^+) \to \pi_q(BESA_1^+) \to \ldots$$

This simplifies as follows. First (12.1) and (7.8) give $K_q R_f$, $K_q SA$ and $K_q SA_1$ as substitutes for $\pi_q(B\bar{E}^+)$, $\pi_q(BESA^+)$ and $\pi_q(BESA_1^+)$ when $q \geqslant 2$. To handle cases with $q < 2$, we ought, in the light of (3.3), (11.2), to consider R_{Sf}, $R_{S^2 f}$, etc. Since $S\mathbb{Z}$ is torsion-free, we can safely tensor by $S\mathbb{Z}$ (over $\mathbb{Z}$), to obtain $S\mathbb{Z} \otimes mA_1$ as the kernel of the epimorphism $S\mathbb{Z} \otimes R_f \twoheadrightarrow S^2A$. (Recall that mA_1 is the kernel of $CA_1 \twoheadrightarrow SA_1$, and thence also of the induced epimorphism

$R_f \twoheadrightarrow SA$.) Therefore SR_f may be identified with the pulled-back ring $(S\mathbb{Z} \otimes CA_1) \times_{S^2A_1} S^2A$ (which also maps onto S^2A with kernel SmA_1). Because of the standard $\mathbb{Z}$-algebra isomorphism $S\mathbb{Z} \otimes C\mathbb{Z} \cong C\mathbb{Z} \otimes S\mathbb{Z}$, this last ring is naturally isomorphic to $(C\mathbb{Z} \otimes SA_1) \times_{S^2A_1} S^2A = R_{Sf}$.

This calculation means we may suspend f as many times as we like before resorting to the above homotopy exact sequence. The upshot may be summarized as follows.

(12.2) THEOREM: *A ring homomorphism* $f : A \to A_1$ *gives rise to a natural exact sequence*

$$\ldots \to K_qA_1 \longrightarrow K_qR_f \longrightarrow K_{q-1}A \xrightarrow{f_*} K_{q-1}A_1 \to \ldots$$

for all $q \in \mathbb{Z}$, *where* $R_f = CA_1 \times_{SA_1} SA$.

Needless to say, the low-dimensional cases of this result pre-date the invention of the higher K-theory. The necessary reconciliation is discussed in [**40**]. An example where R_f has effectively been computed is A_1 as field of fractions of a *Dedekind domain* A. (Recall that this means that every ideal of A is equal to a finite product of prime ideals of A.) Then in [**41** pp.85–147], Quillen shows (by very different arguments) that in (12.2) the term K_qR_f may be replaced by $\bigsqcup_m K_{q-1}(A/m)$, where m ranges over the maximal ideals of A. This case is further dealt with by Gersten in [**41** pp.211–243], Keating [**23**] and Sherman [**34**], who obtain splittings and/or short exact sequences within the long exact sequence under various hypotheses.

THE TRANSFER MAP

Under favourable circumstances a ring homomorphism $f : A \to A_1$ may give rise to a homomorphism of K-groups "going the other way", the *transfer map* $f^t : K_iA_1 \to K_iA$. The inspiration for this is, in algebra, the co-restriction/induction map to the representation ring of a group from that of a subgroup of finite index, and, topologically, transfer maps between (generalized) cohomology groups of a space and its finite covering. Both these examples admit generalizations, as does the one we consider below, which is evidently analogous to them. (For one such generalization of our case, see Quillen's article [**41** pp.85–147].)

We shall consider the situation where, as an A-module (via f), A_1 is finitely generated and projective. For the purposes of this discussion, as for others involving the structure of algebraic

K-theory, it is convenient to pass to an equivariant form of the category of f.g. projective A-modules considered in Chapter 2. A group G is given and all modules are assumed to be endowed with an A-linear Gaction (making them modules over the group ring A[G]). Morphisms are required to be both A-linear and G-equivalent. The short exact sequence Grothendieck group (formed by equating the isomorphism class [P] of P with [N] + [Q] whenever there is a short exact sequence $N \rightarrowtail P \twoheadrightarrow Q$) will be denoted by $R_A(G)$. (I abandoned the search for a notation which mutually specializes both to the ring $R_k(G)$ of k-representations when A is a field k and also to K_0A when G is trivial.)

Whatever relevance $R_A(G)$ may have is due to the existence of a function $\kappa : R_A(G) \to KA^0(BG)$, which I shall now describe. Note that the target group may also be written as $K_0A \times [BG, BGLA^+]$, where the latter factor equates to $\overline{K}A^0(BG)$. Thus $\kappa = (\kappa_0, \kappa_1)$. The image of κ_0 in K_0A is simply obtained by group restriction, since $R_A(1) = K_0A$. To map into $[BG, BGLA^+]$, we begin by adding a G-trivial f.g. projective A-module Q to a given A[G]-module P so as to make $P \oplus Q$ free (over A). Now an A-linear G-action on a f.g. free A-module corresponds precisely to a representation $\rho : G \to GLA$. The effect of different choices of basis for $P \oplus Q$, and then of representatives for the stable isomorphism classes of P, Q, may be to alter the representation by conjugation (by a not necessarily square matrix, α say, which can be replaced by a square one in GLA by the device of considering $\alpha \oplus \alpha^{-1}$ – *cf.* (1.9)). (Details are spelt out in Milnor's book [31 p.26].) Thus the homotopy class of $B\rho : BG \to BGLA$ is well-defined up to commutators of loops (corresponding to commutators in $GLA = \pi_1(BGLA)$), and these are eliminated, as ever, by the passage to $BGLA^+$. So here $\kappa_1([P]) = q_{BGLA} \circ B\rho$.

(12.3) LEMMA: $\kappa : R_A(G) \to KA^0(BG)$ *is a group homomorphism, natural in* A, G.

Proof. The group restriction map is of course well-behaved. We check that $q_{BGLA} \circ B\rho$ sends a short exact sequence of A[G]-modules to a sum of elements in the abelian group $[BG, BGLA^+]$ (where addition is, after Chapter 1, induced from direct sum of matrices). Again, by adding on complementary G-trivial A-modules as necessary, we reduce to exact sequences $A^n \rightarrowtail A^{2n} \twoheadrightarrow A^n$, for which the representation ρ corresponds to a homomorphism

$$G \to \begin{pmatrix} GL_nA & M_nA \\ 0 & GL_nA \end{pmatrix} \subset GL_{2n}A.$$

On the other hand, addition in $[BG, BGLA^+]$ corresponds to the direct sum

$$G \to \begin{pmatrix} GL_nA & 0 \\ 0 & GL_nA \end{pmatrix} \subset GL_{2n}A .$$

So the result hinges on the relationship between $BGLUT^+$ and $BGL(A\oplus A)^+$. In (5.12) (3) it was shown that the canonical maps between them are homotopy equivalences, whence the two factorizations

$$BG \to BGLUT^+ \to BGLA^+$$

and

$$BG \to BGL(A\oplus A)^+ \to BGLA^+$$

are in the same homotopy class after all.

Let us now return to the transfer, and the map $f : A \to A_1$. First, the act of regarding f.g. projective A_1-modules as f.g. projective A-modules determines the transfer homomorphism $f^t : R_{A_1}(G) \to R_A(G)$ for any group G. We now choose $G = GL_nA$ for each n, and note the existence of a distinguished element, denoted $[GL : A^n]$, in $R_A(GL_nA)$, corresponding to the standard GL_nA-action on A^n. Because $[GL : A^{n+1}] \in R_A(GL_{n+1}A)$ restricts to $[GL : A^n] + [A] \in R_A(GL_nA)$ (trivial action on $[A]$), it is preferable to concentrate on the difference element $[GL : A^n] - n[A] \in R_A(GL_nA)$. The fact that these difference elements all restrict to one another means they define an element $\langle GLA \rangle$ in $\varinjlim R_A(GL_nA)$.

It follows that

$$\kappa_1 \circ f^t(\langle GLA \rangle) \in \varprojlim\, [BGL_nA_1, BGLA^+] ,$$

so that application of the plus-construction yields an element

$$t(f) \in \varprojlim\, [BGL_nA_1^+, BGLA^+] .$$

Since $BGL_nA_1^+ \subset BGL_{n+1}A_1^+$ (after (9.6)), the homotopy extension property for CW-pairs allows choice of maps $t(f)_n : BGL_nA_1^+ \to BGLA^+$ representing t(f) which restrict to one another, so defining (non-uniquely) a map $BGLA_1^+ \to BGLA^+$. Now a map g from a compact CW-complex X to $BGLA_1^+ = \varprojlim BGL_nA_1^+$ must factor through some $BGL_nA_1^+$; by virtue of the above constraints on $t(f)_n$, choice of n is immaterial, and the composite homotopy class $t(f) \circ g$ is well-defined. So composition with t(f) defines the required transfer homomorphism

$$f^t : \overline{K}A_1^0(X) \to \overline{K}A^0(X),$$

and in particular (when X is a sphere)

$$f^t : K_iA_1 \to K_iA .$$

Moreover, $1 \otimes f : S\mathbb{Z} \otimes A \to S\mathbb{Z} \otimes A_1$ makes the latter ring f.g. projective over the former, permitting repetition of the above. On observing from (7.2) that composition with $t(S^m f)$ leads to a well-defined map

$$[BGL_nS^mA_1^+, BGLS^mA^+] \longrightarrow [(BGL_nS^mA_1)^+_{m+1}, BGLS^mA^+_{m+1}] ,$$

we may conclude . . .

(12.4) THEOREM: *Suppose* $f : A \to A_1$ *is a ring homomorphism which makes* A_1 *a finitely generated projective* A-*module. Then, for any compact CW-complex* X *and integer* n, f *determines a natural, contravariant transfer homomorphism of groups*

$$f^t : \overline{K}A_1^n(X) \to \overline{K}A^n(X),$$

as well as

$$f^t : K_nA_1 \to K_nA .$$

An especially simple example concerns a finite field extension $f : k \to k_1$, where the composite

$$f_* \circ f^t : K_ik \to K_ik_1$$

is just multiplication by the degree of the extension. (See Quillen's article, loc. cit.) This is an immediate consequence of the projection, or reciprocity formula, which is in turn an exercise once the multiplicative structure on K-groups has been defined – our next task.

13 λ-ring structure

The objective here is to observe how the spectrum $\bar{K}A$, or equally its associated cohomology theory $\bar{K}A^n(X) = [X, \bar{K}A_{(n)}] = [X, (BGLS^nA)^+_{n+1}]$, enjoys a rich structure by virtue of the tensor product and exterior power constructions. The former leads to a graded multiplication and then the latter to the full λ-ring properties (first studied in connection with representation rings and topological K-theory – see [2], [25]). In particular, these extra data carry over to the graded group $K_*A = \oplus K_nA$.

An exposition of multiplication is presented in Loday [29], with such thoroughness that there is nothing to be gained by repeating it here. Instead I offer the following treatment. Its key idea is to view the tensor product of a p× p-matrix over A with a q× q-matrix over A′ as defining, not a pq× pq-matrix over $A \otimes A'$ (recall $\otimes$ always means $\otimes_{\mathbb{Z}}$), but a p× p-matrix of q× q-matrices over $A \otimes A'$. If $\alpha = (a_{ij}) \in M_pA$ and $\alpha' = (a'_{rs}) \in M_qA'$, then we obtain $a_{ij} \otimes \alpha'$ as the (i, j)-th such matrix, where $a_{ij} \otimes \alpha'$ is a q× q-matrix whose (r, s)-entry is $a_{ij} \otimes a'_{rs} \in A \otimes A'$. That is,

$$(\alpha, \alpha') \quad \leftrightarrow \begin{pmatrix} a_{11} \otimes \alpha' & \dots & a_{1p} \otimes \alpha' \\ \vdots & & \vdots \\ a_{p1} \otimes \alpha' & \dots & a_{pp} \otimes \alpha' \end{pmatrix}, \quad a_{ij} \otimes \alpha' = \begin{pmatrix} a_{ij} \otimes a'_{11} & \dots & a_{ij} \otimes a'_{1q} \\ \vdots & & \vdots \\ a_{ij} \otimes a'_{q1} & \dots & a_{ij} \otimes a'_{qq} \end{pmatrix}$$

(In other words, we stop short of identifying $A^p \otimes A'^q$ with $(A \otimes A')^{pq}$.)

This defines a (functorial) ring homomorphism

$$M_pA \otimes M_qA' \longrightarrow M_p(M_q(A \otimes A')),$$

which is compatible with the inclusions $M_pA \hookrightarrow M_{p+k}A$, $M_qA' \hookrightarrow M_{q+k}A'$ given by direct sum with the k× k-zero matrix. So there is no difficulty in passing to the limits mA, mA′ respectively, the pseudo-rings of finite matrices over A, A′, discussed in Chapter 1. The resulting pseudo-ring homomorphism

$$mA \otimes mA' \to m(m(A \otimes A'))$$

composes with the natural isomorphism

$$\gamma^* : m(m(A \otimes A')) \to m(A \otimes A')$$

of (1.13) to yield a natural pseudo-ring homomorphism

$$mA \otimes mA' \to m(A \otimes A').$$

Again using the techniques of Chapter 1, we pass to the associated monoids, where restriction to invertible elements determines a group homomorphism

$$\mu : (mA \otimes mA')^* \to GL(A \otimes A').$$

To finish the construction, we need only (!) exhibit a monoid homomorphism $mA \times mA \to mA \otimes mA$, so that restriction to invertibles gives the required group homomorphism $GLA \times GLA \to GL(A \otimes A')$, all ready for the classifying-space functor and plus-construction. Alas, the lack of a multiplicative identity in mA, mA' puts paid to this attempt at elegance. As a substitute, we define monoid homomorphisms, for each $k, p, q \in \mathbb{N}$,

$$\begin{aligned} \nu^k_{p,q} : M_pA \times M_qA' &\to mA \otimes mA' \\ (\alpha, \alpha') &\mapsto \alpha \otimes I_{q+k} + I_{p+k} \otimes \alpha' + \alpha \otimes \alpha' \\ &= (\alpha \otimes I_{q+k}) \star (I_{p+k} \otimes \alpha'). \end{aligned}$$

Clearly $\nu^k_{p+1,q+1}$ restricts to $\nu^{k+1}_{p,q}$; also

$$\nu^{k+1}_{p,q}(\alpha, \alpha') \star \nu^k_{p,q}(\alpha,0) \star \nu^k_{p,q}(0,\alpha') = \nu^k_{p,q}(\alpha, \alpha') \star \nu^{k+1}_{p,q}(\alpha,0) \star \nu^{k+1}_{p,q}(0, \alpha'). \qquad (\bigstar)$$

Restriction to invertibles, and composition with μ, produces natural group homomorphisms

$$\mu^k_{p,q} : GL_pA \times GL_qA' \to GL(A \otimes A'),$$

leading (thanks to (5.6)) to

$$B\mu^{k\,+}_{p,q} : BGL_pA^+ \times BGL_qA'^+ \to BGL(A \otimes A')^+.$$

Now write $V_p = BGL_pA^+$, $W_q = BGL_qA'^+$. To obtain a map factoring through $V_p \wedge W_q$, we take a representative for

$$[B\mu^{k\ +}_{p,q}] - [B\mu^{k\ +}_{p,q} \circ in_V] - [B\mu^{k\ +}_{p,q} \circ in_W]$$

in the abelian group

$$[V_p \times W_q, BGL(A \otimes A')^+] = \overline{K}A \otimes A'(V_p \times W_q)$$

(where $in_V : V_p \hookrightarrow V_p \times W_q$, etc.). Any such representative must be nulhomotopic on $V_p \vee W_q$. By using the homotopy extension property for the CW-pair $(V_p \times W_q, V_p \vee W_q)$, we can find one such which is actually trivial there, and thereby factors through the quotient space $V_p \wedge W_q$ as desired, yielding $\theta^k_{p,q} : V_p \wedge W_q \to BGL(A \otimes A')^+$. Now one may deduce from (★) that the homotopy class of $\theta^k_{p,q}$ is independent of k. Then, because $\nu^k_{p+1,q+1}\big| = \nu^{k+1}_{p,q}$, so too $\theta^h_{p+1,q+1}\big| \simeq \theta^{k+1}_{p,q}$. In this way, a natural class of maps

$$\theta : BGLA^+ \wedge BGLA'^+ \to BGLA \otimes A'^+$$

may be defined, whose restriction to each $V_p \wedge W_q$ is a unique homotopy class.

The next stage of the project is routine verification of bilinear, associative and commutative properties, and even independence of the class of θ from the only real choice we have made, namely the bijection $\gamma : \mathbb{N} \to \mathbb{N} \times \mathbb{N}$. A crucial lemma here is to be found in [**29**]. (Note that $\delta_{r,s}$ = Kronecker delta.)

(13.1) LEMMA: *Let* $u : \mathbb{N} \to \mathbb{N}$ *be an injection, inducing* $u_\bullet : GLA \to GLA$ *by means of*

$$u_\bullet(\alpha)_{r,s} = \delta_{r,s} \qquad \text{if } (r, s) \notin u(\mathbb{N}) \times u(\mathbb{N}),$$

$$u_\bullet(\alpha)_{u(i),u(j)} = \alpha_{i,j}\,.$$

Then $Bu_\bullet^+ \simeq id : BGLA^+ \to BGLA^+$.

Finally, (10.7) applies.

(13.2) THEOREM: *There is a natural product of spectra*

$$\overline{K}A_{(m)} \wedge \overline{K}A'_{(n)} \to \overline{K}A \otimes A'_{(m+n)}\,,$$

which induces a bilinear, associative multiplication

$$\overline{K}A^m(X) \times \overline{K}A^n(Y) \to \overline{K}A \otimes A'^{m+n}(X \wedge Y)\,.$$

The first specific instance of (13.2) comes when X, Y are spheres.

(13.3) COROLLARY: Theorem 13.2 *defines a multiplication for* $i, j \geqslant 1$

$$K_iA \times K_jA' \to K_{i+j}A \otimes A'\,.$$

The restriction to $i, j \geqslant 1$ in (13.3) is unnecessary, the multiplication extending to lower K-groups by consideration of suspended rings, after (3.3). As a special case, let $A' = S\mathbb{Z}$, $j = 1$. Then $K_1 S\mathbb{Z} \cong K_0\mathbb{Z}$ (3.3) b) is infinite cyclic (Chapter 2). After (3.4), its generator is the class of the infinite matrix $(\delta_{r,s+1}) \in GL_1 S\mathbb{Z}$, coming from $t \in \mathbb{Z}[t, t^{-1}]^* \subset K_1\mathbb{Z}[t, t^{-1}]$. From bilinearity, the multiplication

$$K_iA \times K_1 S\mathbb{Z} \longrightarrow K_{i+1} SA$$

is completely determined by the map

$$\cdot(\delta_{r,s+1}) : K_iA \to K_{i+1} SA\,.$$

Since the two groups are known to be isomorphic (11.2), this homomorphism certainly has potential. After a detailed homotopy-theoretic investigation of the connecting homomorphism ∂ in the exact homotopy sequence of the fibration

$$BGLA^+ \longrightarrow BGLCA^+ \longrightarrow BESA^+$$

studied in Chapters 10, 11 above, namely $(i \geqslant 2)$

$$\ldots \longrightarrow K_{i+1} SA \xrightarrow{\partial} K_iA \longrightarrow K_iCA\ (=0) \longrightarrow K_iSA \longrightarrow \ldots,$$

Lodey in [**29**] drew the following conclusion.

(13.4) PROPOSITION: *The homomorphism*

$$\cdot(\delta_{r,s+1}) : K_iA \longrightarrow K_{i+1} SA$$

is inverse to the isomorphism

$$\partial : K_{i+1}SA \to K_1A \, .$$

Now there is a factorization

$$\begin{array}{ccccccc} K_iA & \xrightarrow{(id,\, t)} & K_iA \times K_1\mathbb{Z}[t, t^{-1}] & \xrightarrow{\cdot} & K_{i+1}A[t, t^{-1}] & & \\ & & \downarrow & & \downarrow & & \\ & & K_iA \times K_1 S\mathbb{Z} & \xrightarrow{\cdot} & K_{i+1}SA & \xrightarrow{\partial} & K_iA \end{array}$$

So a corollary of (13.4) is that

$$\partial : K_{i+1}A[t, t^{-1}] \longrightarrow K_iA$$

is a split epimorphism. This result is strongly suggestive of (3.4), and in fact the proof of the fundamental theorem is completed by Grayson (after Quillen) in [**43**].

(13.5) THEOREM: Theorem 3.4 *holds for all integers* n.

Again, for regular rings more is true. (Compare (3.6).) The proof is to be found in Quillen's article [**41** pp.85–147].

(13.6) *If* A *is left regular, then for all integers* q

$$K_qA \longrightarrow K_qA[t]$$

is an isomorphism.

The next specialization of (13.2) requires that A be *commutative*, which hypothesis we shall assume for the remainder of the chapter. Composition of the multiplication of (13.2) with the homomorphism induced from the A-multiplication $A \otimes A \to A$ yields a multiplication

$$\overline{K}A^m(X) \times \overline{K}A^n(Y) \to \overline{K}A^{m+n}(X \wedge Y) \, .$$

This may be developed in one (only) of two ways.

First, when X, Y are spheres it reduces to a *graded ring* structure

$$K_iA \times K_jA \to K_{i+j}A$$

on K_*A (with extension to lower groups again via suspension) bearing all the standard properties – naturality, associativity and commutativity (often called anti-commutativity because, for $x \in K_i A$, $y \in K_j A$,

$$x \cdot y = (-1)^{ij}\, y \cdot x \quad).$$

Secondly, if $X = Y$, then there is available the diagonal embedding $X \to X \wedge X$, giving rise to a graded ring structure on $\overline{K}A^*(X)$. Note that one cannot have this both ways, for if $X = Y = S^n$ $(n \geqslant 1)$, then the diagonal map $S^n \to S^{2n}$ must be nulhomotopic, so that the product evanesces.

λ-OPERATIONS

Just as with the transfer map discussed in the previous chapter, it is probably simpler to think of exterior power operations in terms of modules rather than matrices. So again (following [**26**]) we perform the construction on the representation ring $R_A(G)$, and then pass to algebraic K-theory by watching its effect on the distinguished element $\langle GLA \rangle$ of $\varprojlim R_A(GL_n A)$. (Remember that A is to be commutative.)

In the usual manner, the k-th exterior power $\Lambda^k M$ $(k \geqslant 1)$ of an A-module M is defined as the quotient of the k-th tensor power $M \otimes_A \ldots \otimes_A M$ by the submodule generated by elements of the form $m_1 \otimes \ldots \otimes m_k$ with some $m_i = m_j$ for $i \neq j$. The A-linear G-action on M determines another on $M \otimes_A \ldots \otimes_A M$, and so restricts to one on $\Lambda^k M$. As a result, its isomorphism class is a well-defined element, called $\lambda^k([M])$, in $R_A(G)$.

Now the function $\lambda^k : R_A(G) \to R_A(G)$ defined thereby is not in general a group homomorphism; indeed, one recalls the A-module isomorphism

$$\Lambda^k P \cong \bigoplus_{i+j=k} \Lambda^i N \otimes_A \Lambda^j Q$$

(leading to the corresponding expression in $R_A(1) = K_0 A$) which comes from a short exact A[G]-module sequence $N \rightarrowtail P \twoheadrightarrow Q$. (One takes $\Lambda^0 M$ to be the trivial A[G]-module A.) Tensoring over A evidently defines a product structure on $R_A(G)$ (and may be employed as for the transfer in Chapter 12 in order to define that on **KA**), suggesting the formula

$$\lambda^k[P] = \sum_{i+j=k} \lambda^i[N] \cdot \lambda^j[Q] \;.$$

This is verified by checking the existence of an A[G]-module map

$$\Lambda^i N \otimes_A \Lambda^j Q \longrightarrow M_i/M_{i+1} \ ,$$

where $M_i \subset \Lambda^k P$ is the submodule generated by products $p_1 \wedge \ldots \wedge p_k$ with at least i terms in N. As this map is an A-module isomorphism, it is an A[G]-module isomorphism too; whence the formula follows from the short exact sequences

$$M_{i+1} \rightarrowtail M_i \twoheadrightarrow M_i/M_{i+1},$$

owing to $M_0 = \Lambda^k P$ and $M_k = \Lambda^k N$. Write

$$\lambda_t(x) = 1 + \sum_{k \geq 1} \lambda^k(x)\, t^k \ ,$$

so as to define a function

$$\lambda_t : R_A(G) \longrightarrow R_A(G)[[t]] \ .$$

The fact that, with "addition" in $R_A(G)[[t]]$ given by power series multiplication, λ_t is a group homomorphism (and $\lambda^1 = id$) makes $R_A(G)$ a (pre-) λ-ring; moreover, an appropriate multiplication in $R_A(G)[[t]]$ (sending, for instance, (1+xt) and (1+yt) to (1+x.yt)) makes λ_t a ring homomorphism. A further polynomial identity for $\lambda^h \circ \lambda^k$ in terms of $\lambda^1, \ldots, \lambda^{hk}$ reveals $R_A(G)$ to be a fully-fledged *λ-ring* (sometimes called a "special" λ-ring). In particular, this settles the λ-ring structure of $K_0 A$.

We now proceed similarly to Chapter 12: attack $\langle GLA \rangle \in \varprojlim R_A(GL_n A)$ with λ^k and pass, via κ_1, to maps in $BGLA^+$. Then the plus-construction leads to a well-defined element

$$\Lambda^k \in \varprojlim \ [BGL_n A^+, BGLA^+] \ ,$$

with $(\Lambda^k)_{n+1}$ restricting to $(\Lambda^k)_n$. Then, for compact X, the function

$$\lambda^k : \overline{K}A^0(X) \rightarrow \overline{K}A^0(X)$$

is defined by sending a class in $[X, BGLA^+]$ represented by

$$i_n^+ \circ f : X \rightarrow BGL_n A^+ \hookrightarrow BGLA^+$$

to that represented by

$$(\Lambda^k)_n \circ f : X \longrightarrow BGL_nA^+ \longrightarrow BGLA^+ \ .$$

This is clearly immune to choice of n. Finally, from the fact that $R_A(GL_mA \times GL_nA)$ is also a λ-ring, one can verify that the functions λ^k on $KA^0(X)$ make $KA^0(X)$ into a λ-ring too. An alternative argument here, reproduced in [**18**], relies on the observation that any function $R_A(GL_nA) \to [BGL_nA, BGLA^+]$ factors uniquely through κ_1. In particular, all this describes K_iA $(i \geqslant 0)$ as a *graded* λ*-ring*. Note that, for $i \geqslant 1$, $\lambda^k : K_iA \to K_iA$ is actually a group homomorphism, for we have seen that all products in K_iA vanish.

There is a standard device for constructing λ-ring homomorphisms ψ^k $(k \geqslant 1)$ (*Adams' operations*) from the λ-ring structure. Let $\psi^1 = id$, and thence, iteratively, for $x \in KA^0(X)$,

$$\psi^k(x) \ = \ \sum_{i=1}^{k-1} (-1)^{i+1}\psi^{k-i}(x) \,.\, \lambda^i(x) + (-1)^{k+1}k\lambda^k(x) \ .$$

Thus on K_iA, $i \geqslant 1$, $\psi^k = (-1)^{k+1}k\lambda^k$, as all product terms vanish. Additionally, $\psi^h \circ \psi^k = \psi^{hk}$, while for p prime, $x \in KA^0(X)$,

$$\psi^{p^r}(x) \ \equiv \ x^{p^r} \ (\mathrm{mod}\ p) \ .$$

These features tend to make the Adams operations more congenial to handle than their λ^k antecedents. Thus [**26**], for a ring A of characteristic $p > 0$, ψ^p behaves as the map on $KA^0(X)$ induced from the *Frobenius* (p-th power) homomorphism. As an application, suppose that A is *perfect*, that is, that $Frob : a \mapsto a^p$ is an automorphism. Then the vanishing of products in K_iA, $i \geqslant 1$, means that λ^p is a homomorphism, and so that

$$\lambda^p \circ p_\bullet \ = \ p_\bullet \circ \lambda^p \ = \ (-1)^{p+1}\psi^p \ ,$$

which is now an automorphism. Hence λ^p and $p_\bullet$ are each automorphisms as well; in particular K_iA is uniquely p-divisible.

Recall from (9.10) that the finite field $\mathbb{F}_q$ has $K_{2j}\,\mathbb{F}_q = 0$ and $K_{2j-1}\,\mathbb{F}_q = \mathbb{Z}/(q^j-1)\mathbb{Z}$, $i \geqslant 1$. It can be shown that $\psi^k = k^j \,.\, id$ on $K_{2j-1}\mathbb{F}_q$. The proof (presented in [**18**], [**26**]) relies on the like behaviour of topological $\psi^k : BGL\mathbb{C} \to BGL\mathbb{C}$, inasmuch as

$$\psi^k = k^j . id : \pi_{2j}(BGL\mathbb{C}) \longrightarrow \pi_{2j}(BGL\mathbb{C}),$$

the key fact (due to Quillen [**33**]) being that $BGL\mathbb{F}_q^+$ is the homotopy fibre of $\psi^q - id : BGL\mathbb{C} \to BGL\mathbb{C}$. Since this may be regarded equally as a deep result about either algebraic or topological K-theory, it seems an appropriate note on which to conclude.

References

1 M.F. Atiyah, "K-Theory", Benjamin (New York, 1967).

2 M.F. Atiyah and D.O. Tall, "Group representations, λ-rings and the J-homomorphism", *Topology* **8** (1969), 253–297.

3 H. Bass, "Algebraic K-Theory", Benjamin (New York, 1968).

4 A.J. Berrick, "The Samelson ex-product", *Quart. J. Math. Oxford* (2) **27** (1976), 173–180.

5 A. Borel, "Cohomologie réelle stable de groupes S-arithmétiques classiques", *C.R. Acad. Sci.* **274** (1972), A–1700–1702.

6 A.K. Bousfield and D.M. Kan, "Homotopy Limits, Completions and Localizations", Lecture Notes in Math. 304, Springer (Berlin, 1972).

7 W. Browder, "Manifolds and homotopy theory", in Lecture Notes in Math. 197, Springer (Berlin, 1971), 17–35.

8 W. Browder, "Algebraic K-theory with coefficients $\underset{\sim}{Z}/p$", in Lecture Notes in Math. 657, Springer (Berlin, 1978), 40–85.

9 A. Dold, "Partitions of unity in the theory of fibrations", *Annals of Math.* **78** (1963), 223–255.

10 A. Dress, "Zur Spectralsequenz von Faserungen", *Inventiones Math.* **3** (1967), 172–178.

11 E. Dror, "A generalization of the Whitehead theorem", in Lecture Notes in Math. 249, Springer (Berlin, 1971), 13–22.

12 E. Dror, "Acyclic spaces", *Topology* **11** (1972), 339–348.

13 Z. Fiedorowicz, "A note on the spectra of algebraic K-theory", *Topology* **16** (1977), 417–422.

14 S.M. Gersten, "K_3 of a ring is H_3 of the Steinberg group", *Proc. Amer. Math. Soc.* **37** (1973), 366–368.

15 M. Hall, "The Theory of Groups", Macmillan (New York, 1959).

16 B. Harris and G. Segal, "K_i groups of rings of algebraic integers", *Annals of Math.* **101** (1975), 20–33.

17 J.-C. Hausmann, "Manifolds with a given homology and fundamental group", *Comment. Math. Helvetici* **53** (1978), 113–134.

18 H.L. Hiller, "λ-Rings and algebraic K-theory", *J. Pure and Applied Algebra* **20** (1981), 241–266.

19 P. Hilton, "On G-spaces", *Bol. Soc. Bras. Mat.* 7 (1976), 65–73.

20 P. Hilton, G. Mislin and J. Roitberg, "Localization of Nilpotent Groups and Spaces", Mathematics Studies 15, North-Holland (Amsterdam, 1975).

21 P. Hilton and J. Roitberg, "On the Zeeman comparison theorem for the homology of quasi-nilpotent fibrations", *Quart. J. Math. Oxford* (2) **27** (1976), 433–444.

22 P. Hilton, J. Roitberg and D. Singer, "On G-spaces, Serre classes and G-nilpotency", *Math. Proc. Camb. Phil. Soc.* 84 (1978), 443–454.

23 M.E. Keating, "A transfer map in K-theory", *J. London Math. Soc.* (2) **16** (1977), 38–42, and **18** (1978), 14.

24 M.A. Kervaire, "Multiplicateurs de Schur et K-théorie", in "Essays on Topology and Related Topics" (Mémoires dediés à Georges de Rham), Springer (Berlin, 1970), 212–225.

25 D. Knutson, "λ-Rings and the Representation Theory of the Symmetric Group", Lecture Notes in Math. 308, Springer (Berlin, 1973).

26 C. Kratzer, "λ-Structure en K-théorie algébrique", *Comment. Math. Helvetici* **55** (1980), 233–254.

27 R. Lee and R.H. Sczczarba, "The group $K_3(\mathbb{Z})$ is cyclic of order forty-eight", *Annals of Math.* **104** (1976), 31–60.

28 R. Lee and R.H. Sczczarba, "On the torsion in $K_4(\mathbb{Z})$ and $K_5(\mathbb{Z})$", *Duke Math. J.* **45** (1978), 101–129.

29 J.-L. Loday, "K-Théorie algébrique et représentations de groupes", *Ann. Scient. Ec. Norm. Sup.* (4e) **9** (1976), 309–377.

30 J.P. May and R. Thomason, "The uniqueness of infinite loop space machines", *Topology* **17** (1978), 205–224.

31 J. Milnor, "Introduction to Algebraic K-Theory", Annals of Math. Studies 72, Princeton Univ. Press (Princeton, 1971).

32 J. Milnor and J.C. Moore, "On the structure of Hopf algebras", *Annals of Math.* **81** (1965), 211–214.

33 D. Quillen, "On the cohomology and K-theory of the general linear group over a finite field", *Annals of Math.* **96** (1972), 552–586.

34 C.C. Sherman, "Some splitting results in the K-theory of rings", *Amer. J. Math.* **101** (1979), 251–295.

35 D. Sjerve, "Homology spheres which are covered by spheres", *J. London Math. Soc.* (2) **6** (1973), 333–336.

36 C. Soulé, "Addendum to the article 'On the torsion in $K_*(\mathbb{Z})$' ", *Duke Math. J.* **45** (1978), 131–132.

37 C. Soulé, "K-théorie des anneaux d'entiers de corps de nombres et cohomologie étale", *Inventiones Math.* **55** (1979), 251–295.

38 E.H. Spanier, "Algebraic Topology", McGraw-Hill (New York, 1966).

39 R.G. Swan, "Excision in algebraic K-theory", *J. Pure and Applied Algebra* **1** (1971), 221–252.

40 J.B. Wagoner, "Delooping classifying spaces in algebraic K-theory", *Topology* **11** (1972), 349–370.

Collected volumes appearing as Lecture Notes in Math., Springer (Berlin):–

41 341 (1973);

42 342 (1973);

43 551 (1976).

Further relevant Lecture Notes in Math., Springer (Berlin), *not cited in text*:–

76 (1968); 108 (1969); 149 (1970); 343 (1973); 734 (1979); 854 (1981);

to appear – ed. K. Dennis.

Index

For topics frequently mentioned, definitions or initial references only are located.